PENGUIN
TOUCHIN

S.K. Das retired from the Indian Administrative Service as Secretary to the Government of India. He completed his education from the University of Delhi and the University of Hawaii, and has been a Visiting Fellow at the University of Oxford. He is the author of *Civil Service Reform* (1998), *Public Office, Private Interest* (2001) and *Rethinking Public Accounting* (2006), all published by Oxford University Press. S.K. Das lives in Bangalore with his wife and two sons.

TOUCHING LIVES

The Little Known Triumphs of the Indian Space Programme

S.K. DAS

PENGUIN BOOKS

PENGUIN BOOKS
Published by the Penguin Group
Penguin Books India Pvt. Ltd, 11 Community Centre, Panchsheel Park,
New Delhi 110 017, India
Penguin Group (USA) Inc., 375 Hudson Street, New York, New York 10014, USA
Penguin Group (Canada), 90 Eglinton Avenue East, Suite 700, Toronto,
Ontario, M4P 2Y3, Canada (a division of Pearson Penguin Canada Inc.)
Penguin Books Ltd, 80 Strand, London WC2R 0RL, England
Penguin Ireland, 25 St Stephen's Green, Dublin 2, Ireland
(a division of Penguin Books Ltd)
Penguin Group (Australia), 250 Camberwell Road, Camberwell,
Victoria 3124, Australia (a division of Pearson Australia Group Pty Ltd)
Penguin Group (NZ), 67 Apollo Drive, Rosedale, North Shore 0632,
New Zealand (a division of Pearson New Zealand Ltd)
Penguin Group (South Africa) (Pty) Ltd, 24 Sturdee Avenue, Rosebank,
Johannesburg 2196, South Africa

Penguin Books Ltd, Registered Offices: 80 Strand, London WC2R 0RL, England

First published by Penguin Books India 2007

Copyright © S.K. Das 2007

All rights reserved

10 9 8 7 6 5 4 3 2 1

The views and opinions expressed in this book are the author's own and the facts are as reported by him/her which have been verified to the extent possible, and the publishers are not in any way liable for the same.

ISBN 10: 0-14310-216-8   ISBN 13: 978-0-14310-216-8

Typeset in *Sabon Roman* by SŪRYA, New Delhi
Printed at Thomson Press (India) Ltd.

This book is sold subject to the condition that it shall not, by way of trade or otherwise, be lent, resold, hired out, or otherwise circulated without the publisher's prior written consent in any form of binding or cover other than that in which it is published and without a similar condition including this condition being imposed on the subsequent purchaser and without limiting the rights under copyright reserved above, no part of this publication may be reproduced, stored in or introduced into a retrieval system, or transmitted in any form or by any means (electronic, mechanical, photocopying, recording or otherwise), without the prior written permission of both the copyright owner and the above-mentioned publisher of this book.

For the ISRO family—past and present—who have striven to conquer space and harness its potential for the common man.

Contents

	ACKNOWLEDGEMENTS	IX
	INTRODUCTION	XI
1.	IN ALIRAJPUR: TV TRAVAILS AND DEVELOPMENT COMMUNICATION	1
2.	IN KORAPUT: FINDING DRINKING WATER	25
3.	IN CHAMARAJANAGAR: EDUCATING CHILDREN	40
4.	LAKSHADWEEP: HELPING FISHERMEN	60
5.	IN TRIPURA: PROVIDING HEALTHCARE	77
6.	IN ORISSA: HELPING FLOOD EFFORTS	100
7.	KARNATAKA'S PANCHAYATS: EMPOWERING GRASS-ROOTS REPRESENTATIVES	117
8.	IN JHABUA: WATERSHED	127
9.	THE HIMALAYAS IN GARHWAL: PILGRIMS' PROGRESS	160
10.	THE SUNDARBANS: PROTECTING THE ECOSYSTEM	172
11.	IN THE GANGETIC PLAINS: RECLAIMING SODIC LAND	189
12.	IN GOD'S OWN COUNTRY: AUTOMATIC WEATHER STATION	212
13.	IN THYAGARAJA'S LAND: THE VILLAGE RESOURCE CENTRE	231
	INDEX	252

Acknowledgements

I received much generous assistance from a large number of people in the course of writing this book. Among those who were closely involved, several names stand out.

I thank Dr K. Kasturirangan, G. Madhavan Nair, Dr R.R. Navalgund, Dr K. Radhakrishnan, S. Prabhakaran, V. Sundararamaiah, A. Bhaskaranarayana, Dr V. Jayaraman, Dr V.S. Hegde, S. Krishnamurthy, L.S. Satyamurthy, Dr S.V. Kibe, Dr Shailesh Nayak, G. Behera, P.G. Diwakar, Dr A. Jeyaram, B. Manikiam, S.K. Srivastava and S. Bandopadyaya. I am indebted to all of them.

I thank Ravi Singh and Avanija Sundaramurti of Penguin Books India for their help and support. They have been good to me in countless ways, and this despite the fact that it must be particularly trying to put up with someone from the government writing a book on the government.

It is especially pleasant to acknowledge and thank the members of my wonderful family for their cooperation and encouragement to complete the book on time.

Introduction

The Indian space programme was born in a church. The church is like any other that dots Kerala's verdant landscape, but it has a hoary past—it was founded by Saint Francis Xavier in 1544. To start with, the church was a shed thatched with dried coconut leaves. The fisherfolk from the coastal village of Thumba used to gather in this shed to listen to the Bible teachers. Once a year, a priest from Trivandrum would visit the church to administer sacraments.

The space programme started with launching of sounding rockets. The idea was to take up scientific investigations of the unique upper atmospheric and ionospheric phenomena above the geomagnetic equator. Since this passed through Thumba village, it was necessary to locate the programme at Thumba. The only suitable building available in the village was the church and so the programme was started there.

The first sounding rocket took-off in November 1963. Sounding rockets were nothing new at the time; countries in the West had been launching them for almost two decades by then. But the zeal with which ISRO's engineers built and flew those rockets was missionary; much like the zeal of the Bible teachers in the sixteenth century who braved the elements to preach in the thatched church.

Subsequently, ISRO's engineers built and launched more

sophisticated rockets with the same missionary zeal. The big break came with the Polar Satellite Launch Vehicle (PSLV) that became ISRO's workhorse for launching remote sensing satellites. But even with an operational PSLV, there was something missing: ISRO had no muscle to climb to 36,000 kilometres carrying its communication satellites. Geo-Synchronous Satellite Launch Vehicle (GSLV) did that in 2001, making India the sixth country in the world to have that capability.

ISRO's record in building satellites was equally impressive. Starting with IRS-1A, ISRO has built and launched quite a few remote sensing satellites; they are easily the most versatile civilian satellites for looking at Earth and its oceans. ISRO has also built a large number of satellites for communication and meteorology; it now has one of the biggest communication satellite systems in the world, its reach extending from Europe to Australia to Japan.

ISRO has capabilities that are comparable to the best anywhere in the world, but what makes it different is the way in which ISRO's satellites deliver services to the society, by shrinking both time and distance. The broadcasting capabilities of ISRO's satellites, for example, are used not only for entertainment, but also for tele-education, telemedicine, training and development communication. Using its satellites, ISRO has virtually taken the best teachers in the country to teach students in remote villages who are not lucky enough to be in cities. It has linked city hospitals to health centres in rural areas and tiny towns, bringing to them the skills of specialists who work only in the big cities. ISRO has made it possible for experts sitting in district or state headquarters to interact on socially relevant themes with villagers and extension workers in rural areas.

ISRO's satellites have been especially useful at the time of disasters. Through receivers located in disaster-prone villages, the satellites provide advance warning of impending disasters, helping save thousands of lives and livestock. During floods, ISRO generates inundation maps very quickly, identifying marooned areas to enable rescue and relief. ISRO has created a digital database depicting areas vulnerable to disasters such as cyclone, floods, drought, landslides and earthquakes. The database helps in rapid monitoring and management of natural disasters.

Data from its remote sensing satellites has helped ISRO achieve a number of tasks. It prepares groundwater maps for the entire country. Using these maps, ISRO has located sites for drilling wells in villages that have acute shortage of drinking water. ISRO has mapped forests, wastelands and even habitats that harbour vector mosquitoes spreading the dreaded brain fever. From monitoring irrigation commands, reclaiming waterlogged and salt affected lands, and planning for watershed development in rain-fed rural areas to assessing the environmental impact of development projects, characterizing biodiversity at landscape level, and helping in preparing working plans for afforestation, ISRO's remote sensing data has helped do it all. Even the judiciary used remote sensing date to stop forests from being encroached.

Data from remote sensing satellites has enabled ISRO to assess area under various crops and predict their yield about a month in advance of harvest to a level of accuracy that is as high as 90 per cent. The synoptic vision of a satellite combined with such information services have provided necessary inputs to plan for cropping regimes in different parcels of agricultural land, taking into account agro-climatic characteristics, resource endowments and market conditions. These strategies have helped in sustaining productivity of land and water resources, diversifying agriculture and maximizing returns to the farmers.

ISRO's remote sensing satellites have performed two useful tasks. For one, they have enabled surveys of natural resources so that the information they provide becomes the basis for planning and formulation of policy. Second, when these policies are implemented, the satellites have monitored the results. Data from these satellites has helped us to understand the nuanced relationship among various resources and what human intervention means in terms of its impact. Such an understanding has been the key to judicious use of natural resources and to evolve benign strategies for their sustainable development.

Using the vantage position of space, ISRO has delivered a variety of services. But the question is: Have ISRO's services contributed to improving the life of the ordinary citizen? Is space now a community resource that has outreach to one and all? Has ISRO touched the life of the common man? These are questions this book sets out to discover answers to, based on interactions with ordinary people living in the remotest parts of India.

1

IN ALIRAJPUR
TV Travails and Development Communication

For an eminent scientist and head of an organization like ISRO, Dr Kasturirangan smiles a lot. It is always a quick smile; a smile that starts with a twitch of the right corner of his mouth, and then, a blaze of light from his expressive eyes. The smile is very engaging.

But on the day he sent for me, he was not smiling. There was a deep frown on his face as he gave me a sheet of paper to read. It was a letter from Alirajpur, a small town in Jhabua district in Madhya Pradesh where ISRO was implementing a project. In the project, specially designed TV sets had been installed in villages, broadcasting programmes through an ISRO satellite to reach the villagers every evening.

The letter was anonymous. It said the ISRO TV in Alirajpur block office had been taken away by some officer to his house and the villagers had not seen the broadcasts for quite some time.

Dr Kasturirangan waited for me to finish reading the letter and then said, 'I'm so disappointed! ISRO takes up a programme to help the poor villagers and see what happens!'

'It may not be true, after all,' I took the liberty to suggest. 'People do often make false allegations, sir.'

His frown deepened. And there was a hint of reproach in his voice.

'Well, a serious allegation has been made,' he said. 'The least we can do is to look into it. I'd like you to go there and find out what's going on. Listen, while you're there, you might as well see how our programme is doing in other villages. Talk to the villagers themselves, will you?'

I braced myself for the question I was going to ask.

'Why me, sir?' I asked. 'I am only a finance man. I don't even know what the programme is about!'

He smiled. That quick, engaging smile.

'Why you?' he said evenly. 'Because I want an objective assessment. Take Dr Kibe with you. He knows everything about the programme. One more thing. Not a word to the Jhabua collector about your trip. No intimation to ISRO people in Jhabua either. I want you to go there without any announcement and find out how things are.'

'I'll do that,' I said tamely.

'When do you think you can start?' he asked in a tone that, I knew, would brook no delay.

'I'll start tomorrow', I said.

I was almost out of his room when he called me back to announce, 'One last thing, Sisir. I'd like proper documentation to be made. Take a video photographer with you. That way, I can get an account of what people in Jhabua think of our programme.'

※※※

When we landed at Indore the next day, our first task was to find a taxi that would take us to Alirajpur and other villages in Jhabua district. We were lucky to get a Tata Sumo with a driver who seemed to know Jhabua rather well. He was charming and companionable, but he had a flaw: he was ready to talk his head off at the slightest provocation. He helped us find a video photographer who was rather economical with his expressions. A compensation of sorts, I thought.

We had our lunch at Indore and set off for Alirajpur. I started talking to Dr Kibe.

'I know very little about the Jhabua project,' I said. 'Why don't you tell me something about it?'

'You see,' Dr Kibe explained, 'ISRO has always tried to use its satellites for promoting development. We started the SITE programme in 1975-76. You know what we did? We put 2400 TV sets in isolated villages in six states of India. Through these TV sets, we broadcast programmes to these villages from a satellite.'

'What kind of programmes?'

'Oh, all kinds. About agriculture, animal husbandry, dairy, poultry, health and hygiene, family planning, and education. Some entertainment programmes, too. The programmes were in the local language and broadcast for four hours every day.'

'What was the impact like?'

'Oh, the programmes were received very well. You see, these viewers were illiterate. They were from the poorer sections of the rural society. You know something? They had never read newspapers or gone to a cinema. They hadn't even listened to a radio!'

'How long did the SITE programme go on for?'

'Only for a year. Because the satellite that carried the programme was an American satellite. Available to us only for a year. But we carry the Jhabua programme on ISRO's own satellite.'

'What is the configuration in the Jhabua project?'

'Well, we have an Earth Station in Ahmedabad as the uplink station. We also have a studio there for transmission of programmes. We make the programmes ourselves. We have located Direct Reception Sets in the villages of Jhabua district to receive the programmes from ISRO's satellite.'

'When are the telecasts done?'

'In the evenings. They are for two hours every evening for five days a week. We telecast fifteen minutes of news from Doordarshan Bhopal on all five days. Followed by four Jhabua-specific programmes of about twenty minutes each. Rest of the time, we transmit short programmes such as songs, dances and development quickies.'

'What kind of things do you talk about in these telecasts?'

'About things that have a bearing on the village life in Jhabua: agriculture, animal husbandry, poultry, fisheries, environment, employment, watershed, education, government schemes, forestry, health and hygiene, and panchayati raj.'

'Very comprehensive, isn't it?'

Dr Kibe nodded. 'Yes. There are also programmes on social issues like crime and dowry. Also programmes on social issues which are typical to Jhabua. Like superstitions, excessive expenditure on marriages and death ceremonies, and alcoholism.'

'Who looks after these TV sets in the villages?'

'We have custodians. They are responsible for the operations and for the safe custody of the equipment. We train them to operate the equipment, and report faults, if any.'

'Tell me, why did you choose Jhabua district?'

'For two reasons. The terrain is very hilly in Jhabua. The habitation is scattered. The communication facilities are awful. That makes Jhabua an ideal place for implementing a satellite-based development communication project.'

That's true, I thought. A satellite can transcend geographical barriers.

'And also because of the economic backwardness of the district. It is perhaps the most backward district in the country. And certainly the poorest. Almost 85 per cent of the population in the district is tribal. They are the Bhils, and are very poor. The district has a very low literacy rate, high dropout rate and high infant mortality rate.'

The driver was making grunting noises for quite some time and now he delivered his coup de grâce. In a voice laden with contempt, he said, 'As if those wretched Bhils will have any time for watching these programmes! If they spend their evenings before the community TV, who'll do the looting, the dacoity?'

The final summation delivered, he swerved the Tata Sumo with a flourish to a roadside stall, with the announcement:

'Time for some tea, folks!'

※※※

It seemed as if the tea had revived the driver's sagging spirits, he now drove with great gusto. But my eyes were glazing over, I dozed off peacefully.

I woke up with a jolt. We were being tossed from side to side. The driver was driving really fast and the road was not particularly good. Why is he so reckless?

'What's the hurry?' I asked the driver. 'It doesn't matter if we are a bit late, does it?'

'But what about the wretched Bhils?' he answered my question with a question. 'If we don't reach Alirajpur while there is still some light, don't be surprised if you're waylaid by the Bhils. Looted and beaten up in the bargain!'

That was irrefutable logic. I looked out of the window of the Tata Sumo with a shudder. It was a tangle of gorse and scrub, with an occasional thicket of jungle; ideal territory for an ambush. Were the marauding Bhils lurking in that thicket to waylay us and wreck Dr Kasturirangan's mandate?

'Do the Bhils really do these things?' I asked.

'Of course they do!' our driver said with a sneer. 'All the time. They ambush vehicles on the road and loot passengers. That is fairly common, let me tell you that.'

'Is that true, Dr Kibe?' I asked.

Dr Kibe confirmed that several cases of looting had indeed been reported from Jhabua.

'Well, didn't I tell you?' the driver said gloatingly. 'These Bhils are horrible people. Totally lawless! You know how they do it? They put big boulders on the road and lie in wait behind the hills. When the vehicle stops, they ambush the hapless travellers and loot them.'

'Do these things happen often?' I asked.

'All the time,' the driver replied. 'Let me tell you what these criminal Bhils did the other day on the road from Indore to Ahmedabad. A luxury bus was going at night to Ahmedabad. They put big boulders along the entire width of the road. The bus slowed down and finally came to a stop in front of the boulders. As soon as the bus stopped, they surrounded the bus with their bows and arrows.'

'What happened?' I asked in consternation.

'They relieved the passengers of their cash, clothes and other valuables. You know something? Once it becomes dark, no passing vehicle is safe in Jhabua district. That's why all the vehicles here travel in a convoy with a police escort.'

'Why would the Bhils do such awful things?' I asked.

Dr Kibe intervened to say, 'The Bhils are very poor. Some of them may do such things in sheer desperation.'

There was a big board welcoming us to Jhabua. The landscape changed dramatically. From the even, placid wheat fields, it became undulating—dotted with hills with bald, treeless slopes. Houses, few and far between, stood apart from each other. They were shaped like rectangles, raised above the ground on a layer of Earth and stone; structures of mud, plastered with cow dung and topped by a bamboo thatch. The roof rose at an inclination from the two long sides.

I gestured towards a house and asked, 'That's a Bhil house, isn't it?'

It was the driver who replied.

'Yes, that's a Bhil house.'

'These Bhil houses are kind of scattered,' I observed.

The driver let out a harsh laugh. 'Yes. I'll tell you why. Because the Bhil wants to live as far away from his neighbour as possible. You know the reason? It's because he thinks his neighbour's wife is a witch.'

I was aghast. 'A witch! Don't tell me there are witches in the Bhil villages! Real witches, you mean?'

The driver reacted angrily as if I had cast aspersions on his knowledge of the Bhils. 'Of course, real witches. Let me tell you that there is a witch doctor in each Bhil village. They call him the Badwa. Mind you, he is an important person.'

'Why, what does the Badwa do?'

'Let us say someone falls sick,' the driver explained. 'The Badwa is called. Only he can say why the man is sick. And exorcize the evil influences at work.'

'But the Badwa is not even a qualified doctor!' I protested. 'Or, is he?'

The driver shook his head. 'No, the Badwa is not a doctor. But the Bhils go to him; they'd never go anywhere near a real doctor.'

That's bad, I thought and looked out of the car window. The bald, brown hills kept repeating themselves with a kind of weary insistence. In the clearing between the hills, there were tall trees with green foliage that looked like a well-tended green canopy.

'What's that tall tree?' I asked.

'The mahua tree, sir,' it was the driver who answered. 'That's the tree which makes the Bhils so savage!'

Strange, I thought. A tree with powers of witchcraft?

'It is this mahua tree,' informed our driver, 'which gives the Bhils their liquor. You should be here in April, sir. That's the time the mahua flowers blossom.'

'Why, what happens then?' I asked.

'All hell breaks loose, sir,' the driver declared dramatically. 'The Bhils just go mad! At that time, even the Bhil cattle run amuck. You know something, sir? You have to beat the cattle to push them away from the mahua trees.'

'The cattle getting drunk?' I asked in disbelief. 'That's incredible!'

'Isn't it, sir?' the driver said gloatingly. 'You can see the Bhils—men, women and children—bending under the tree. They gather the mahua fruits and put them away in bags.

You can see men pulling out their arrows and shooting down other men.'

I was shocked. 'Why should they do that?' I asked.

'Because bloody fights break out, sir,' the driver explained. 'Fights over who has the right to collect the magic mahua fruit. And there is a lot of violence.'

'How do the Bhils make the mahua liquor?'

'Simple, really,' the driver explained. 'The Bhils collect the mahua stuff sometimes in April-May. They dry it, then, brew and distil it.'

'How long does it take to make?'

'About eight days.'

'Is mahua potent as a drink?'

'Very potent, sir,' the driver said with a sneer. 'Really knocks you out! There is even a story about it.'

'What story?'

'That the mahua tree grew at a place where a Brahmin, a tiger, a jackal, and a dog were burnt. So when the Bhil drinks mahua, he first speaks in a learned manner like a Brahmin. When he drinks a little more, he becomes a tiger.'

'A tiger?' I asked in amazement.

'Like a tiger, I mean. He quarrels and pulls his moustache. He takes some more mahua, and he is like a jackal. He shouts loudly. And then, some more mahua and he's like a dog. He bites his neighbours, barks abuses at them and then, goes off to sleep.'

It was a joke well-told, but the driver laughed the most at his own joke.

※※※

We entered Alirajpur town. Dusk was settling in and light was fading fast. Dr Kibe asked the driver sternly, 'I hope you know where the block development officer's (BDO) office is, because that's where we are supposed to go.'

The driver's tone was nonchalant. 'Don't worry! I know

this place like the back of my hand. The BDO's office is right opposite the big playground.'

We stopped in front of what seemed to be the town square. 'This is the BDO's office,' the driver announced. 'But there doesn't seem to be a soul inside. I think they have closed for the day.'

'Why don't you go in and find out?' I asked. 'There's bound to be someone inside. At least, a watchman. Ask him about the evening broadcast and the TV set. But please don't tell him who we are.'

'Trust me,' he declared and set off with a swagger. Like a cop in a Hindi movie. When he came back, he was full of news.

'I searched the whole building,' he informed us. 'Do you know where I found him? At the back of the building, drinking tea.'

'Who did you find?' Dr Kibe asked harshly. He obviously didn't like the driver.

'Didn't I tell you?' the driver said in an even tone. 'The watchman, of course. Who else?'

'What did he say?' Dr Kibe's voice was even harsher.

'He said the TV set was there till a few weeks back and there were lots of people every evening to watch ISRO's programmes. Then, a new fellow was posted as the Sub Divisional Magistrate (SDM). He, I believe, is the BDO's big boss. The SDM wanted the TV set in his house. So the set was taken away. Without the TV, people don't come here anymore.'

'So what do we do now?' Dr Kibe asked in a plaintive voice.

'I think we should go in and tell the watchman that we're from Star TV,' it was the video photographer suggesting. 'That we're here to do an item on ISRO's programme. The watchman will go and get the BDO then.'

I looked at him in admiration. So he wasn't as dumb as he looked.

'That's a good idea,' I said. 'But how do we do that? We don't have any identification from Star TV, do we?'

He gave me a toothy smile. 'Well, I've a Star TV identification which I can attach to my equipment,' suggested the photographer. 'I always carry one with me. One never knows when it'll be needed! In the past, it has got me access to places where I couldn't have otherwise gone in a thousand years!'

※※※

We marched off to the BDO's office and told the watchman that we would like to interview the BDO for a programme we were doing for Star TV. He glowered at the driver but went off to fetch the BDO.

It was quite a while before the BDO turned up. He was a man in his early forties, with the furtive look of a crusty bureaucrat. He was dressed in a safari suit that sat a shade too tightly on his corpulence. Our photographer started clicking and the BDO put on what, he thought, was a charming smile.

'If only you had told me beforehand,' he said in his most effusive voice, 'I would have received you with the kind of welcome you folks deserve. Now what government programme do you want me to talk about? The government of Madhya Pradesh has so many excellent programmes for the poor. Such as . . .'

Before he could start on that often-quoted list like a waiter in an Irani restaurant, I cut him off. 'We'd like to interview you about the ISRO programme,' I said. 'How the Bhils have benefited from it.'

'Of course, of course,' he said. 'It's a good programme. But there are other programmes I can tell you about. Such as . . .'

I had to cut him off once again. 'We are here to do a feature on ISRO's programme. We would like to interview you on that. And talk to some Bhils who watch the programme regularly. Also photograph them before the TV.'

'Would that be necessary?' he asked in a voice that

sounded already strained. 'I can tell you all about the programme. Why interview these Bhils? They are very shy people. And why in front of the TV?'

A crowd was beginning to gather around us. It crossed my mind that the flash bulbs of a camera with a Star insignia didn't go off every day in this sleepy town.

'But we'd like to interview you and other people in front of the ISRO TV,' I insisted in my most pleasant voice. 'Why, is there any problem with the TV?'

'Yes,' he said. 'You see, the ISRO TV has gone for repairs.'

I knew he was lying. I said, 'It's such a pity, then. We would've liked to do the programme in Alirajpur. I believe it's such a great success in this block. And we're told that you're the smartest BDO in this part of the world, and we were hoping that we could put you on our programme. Well, looks like it can't be helped. We will go to Kathiwada or Sondwa.'

The BDO looked perturbed. Beads of sweat stood on his forehead that looked like little, unformed pearls in the harsh glare of the camera light. I knew he was caught between the lure of appearing for a national TV channel and the possibility of antagonizing his boss.

'Well, you don't have to go to Kathiwada or Sondwa. You can do the programme here,' he said after a while. 'That TV repair fellow had told me that the TV would be ready today. I'll go and collect it now. You wait for me here.'

He left, accompanied by the watchman. The crowd around us had swelled. They were Bhils. I looked at them with interest. They were of medium height with lean bodies that were tanned brown. They had pleasant faces—broad foreheads, pointed chins, and thin lips. All of them sported enormous moustaches—bushy ones with curved ends. They looked poor. The turbans they wore were dirty and askew, and the long, matted hair hung shabbily to their shoulder.

I thought it was a good time to talk to them. I asked one of them, 'Have you watched any of these programmes on the ISRO TV here?'

He twirled his enormous moustache like it was a plaything and said in an amiable voice, 'Of course, of course. I'm a regular. I like these programmes very much. I used to come here every day. But now the TV is not there.'

'What are the programmes you like?' I asked.

'I like all of them,' he said laconically.

'Anything in particular?' Dr Kibe asked.

He scratched his head as if he was trying to recapture a long lost memory byte. He said after a while, 'A serial called *Koshish*.'

'What's it about?' I asked.

'It's about a schoolteacher. He has just been appointed. He arrives in the Bhil village where he has been posted. He's young. But he finds that the villagers are not interested in education, neither the parents nor the children. But he doesn't give up.'

'What does he do?' Dr Kibe asked.

'He talks to the villagers, the parents and the children. He tells them how important education is. How it'll help them in the future. The teacher also helps the villagers in times of difficulties. That way, he gains the trust of the villagers.'

'Then, what happens?' I asked.

'The villagers start sending their children to the school. The school becomes full of children. I really liked that serial.'

He knit his brow, as if undecided whether to tell us anything more.

'Is there something you want to tell us?' I egged him on.

'All right,' he said, beaming us an uncertain smile. 'After watching the serial, I decided to send my kids to school. Of course, it meant some problems for me. Because the children used to look after the cattle, feed them and fetch water from the stream. Now, I've to do these things myself. Doesn't matter. I'm happy my children are in school.'

He gestured towards the person standing next to him, and said, 'He is Kaliya. He is the one always watching ISRO serials. Why don't you say something, Kaliya?'

'So Kaliya, do you also like this serial *Koshish*?' I asked.

Kaliya spoke so inaudibly that I had to crane my neck to hear him. He said, 'Yes, yes, I like that. But there's another serial I like even more.'

'What's it called?' I asked Kaliya.

'*Aao Akash Ko Chhu Le*. It's a nice serial.'

'Is that so?' I asked. 'What's it about?'

'Oh, about the hopes of a Bhil boy. He wants to do something big in his life. But for that, he has to go to school, study hard. After seeing that serial, I knew that my children must go to school. And study hard.'

'How're they doing in school?' I asked.

A smile lit up his wizened face. 'They are doing well. I now ask them to study hard.'

'Do you also make them do other chores?' I asked. 'Like working in the field or household jobs?'

I waited for Kaliya to answer, but the BDO now marched in, accompanied by the watchman carrying a TV set and another person who walked with a swagger. We followed them to the TV room and so did the crowd. I sat down next to the person with the swagger; he turned out to be Malaviya, the number two man in the BDO's office.

The TV was connected and switched on. It swung into life and the assembled crowd clapped. An episode was just starting, with the title song and a montage. A programme flashed on the screen. It said *Do Kadam Sahi Thor Ki Or*.

'What's that?' I asked Malaviya.

'It means Two Steps in the Right Direction,' he explained. 'It's a very popular programme with the Bhils. It gives information on various government schemes. Schemes for which the Bhils can approach us for financial assistance and advice.'

'How's the serial structured?'

'There are no fixed characters in the serial. No specific script. Each episode begins with a problem that has occurred within a family. Bhils acting the character of other villagers enter the frame and either encourage or discourage the family to seek help. The narrator steps in and poses questions on whether the problem would be solved. Government officials now come into the frame and respond to the problem. They advise on who can avail of the schemes and the addresses of the places from where the benefits would be available.'

The programme started. It was about water and sanitation, and how important they are. It talked of the schemes that provided water and toilet to the Bhils. It also talked about handpumps for drinking water. The programme ended with a round of applause from the crowd in the room and that was the end of the evening's transmission. The Bhils left one by one.

To vindicate our credentials as the roving team from Star TV, we did conduct an interview with the BDO, and another one with Malaviya, duly photographed by the video photographer. While interviewing them, Dr Kibe managed to extract an assurance that the ISRO TV would not be sent for such long, extensive repairs the next time.

※※※

Dr Kibe had drawn up a programme for us to visit a few villages in Jhabua district. Our first destination was Choti Malpur, a village on Alirajpur–Dahod Road. It was a small village with 135 houses and a population of 877. The ISRO TV was kept in the anganwadi building with a certain Mrs Lalita Parmar as the custodian.

Mrs Parmar was waiting for us in the anganwadi building. We found we were in luck. Some of the residents of the village were there with her. Mrs Parmar introduced us to Ratan Bhabar, Karam Singh, Besti Behn and Bhuri Behn. We explained that we were there to find out how ISRO's scheme of development communication was working.

It was Ratan Bhabar who spoke first. 'It's a good scheme.' His voice was gruff. 'So many Bhils come here to watch ISRO's programmes. I'm a regular myself.'

'What are the popular programmes?' I asked.

'All the programmes are popular,' he explained. 'As for me, I like *Ped Baba Ki Kahani*.'

'What is this serial about?' Dr Kibe asked.

'It is the story of the tree god. There is a song that comes every time the serial starts. You know what it is? It is a song the Bhils sing to the tree in real life. We invite the tree to come to our houses as a guest and that every tree is a god.'

'Why, that's very nice!' I said. 'Who are the characters in the serial?'

'One Prema, a Bhil boy. And Dithli, Prema's friend. Then, Prema's grandfather. And also, Prema's sister-in-law.'

'What happens in the serial?' Dr Kibe asked.

'Each programme deals with a problem. Prema and Dithli try to find a solution guided by the Ped Baba, the spirit of an old tree.'

'How's the story narrated?' I asked.

'Well, Ped Baba tells the story. He takes us to Bhil villages where good work has been done by the villagers.'

'What kind of work?' Dr Kibe asked.

'Like growing grass. Or, doing watershed. Building check-dams. Growing forests.'

'Which are these villages?' Dr Kibe asked.

'So many of them. Villages like Hathipawa, Somkuva, Amankuva, Kakradhara, Bhawar Pipliya. You know something? I liked the work that the villagers did in these villages just by coming together. This has helped me in my work.'

'What is your work?' I asked.

'You see, I am the president of the Watershed Development Committee of Choti Malpur village. For that, I have to organize a large number of self-help groups of women. After the Bhils here saw the ISRO programmes and learnt what had been done in other villages, my work became very easy.'

Besti Behn, the president of the self-help group of Choti Malpur village spoke up now.

'What Ratan Bhabar said is true,' she said. 'We've learnt a lot from these ISRO programmes.'

'Like what?' I asked.

'Let me tell you about the Kuldi,' Besti Behn said. 'We've this Kuldi. It means a kind of pot in Bhil language. We keep about Rs 500 in reserve in the Kuldi. Suppose, now, somebody's child were to fall sick. She'll go to the Kuldiwalli's house, take the money from the Kuldi and rush the child to the hospital.'

'That's really good,' I observed. 'But tell me, where did you people get this idea of the Kuldi from?'

'From ISRO's programmes only,' Bhuri Behn declared. 'Where else? There was this programme in *Ped Baba Ki Kahani*. About Bhawar Pipliya village. How the Bhil women there had got together and formed a Bairani Kuldi. We liked that programme very much.'

A singsong voice piped up from the back. 'What about the programme on Rama village? Why don't you tell him about it, Bhuri Behn?'

Bhuri Behn nodded enthusiastically. 'Yes, yes. I forgot. There was this programme in *Ped Baba Ki Kahani* about Rama village. It was about the Bhil women there, how they formed a Bairani Kuldi. With the help of the forest department, they set up a nursery. You know something? They made a profit of Rs 2,00,000. They used the money to give loans to the women of the village.'

It was the singsong voice from the back again. 'We liked that programme so much that on the very next day the women of our village sat down and decided to have a Kuldi. Isn't that so, Bhuri Behn?'

'Yes, yes,' Bhuri Behn declared. 'And we've benefited so much from the Bairani Kuldi!'

Our driver indicated that we must leave now, if we had to keep our appointment in Sanda village. We took leave of Ratan Bhabar, Karam Singh, Besti Behn, Bhuri Behn and the custodian.

※※※

Sanda was a bigger village than Choti Malpur, with 223 houses and a population of 1347. The ISRO TV was kept in the anganwadi building. Mrs Parvati Mohan, the custodian, was waiting for us in the anganwadi building along with two persons. There was Parvati Behn who headed one of the self-help groups in the village, the Shriram group. She had a bright face, with a roguish twinkle in her eyes and an animated expression. The gentleman was Mohan Mavji.

'Do you watch these ISRO programmes?' I asked Mohan Mavji.

'Yes,' he beamed me a wide smile. 'I'm a regular. I haven't missed a single programme so far.'

'Do you like them?' I asked.

'Yes, very much. But there are some programmes I like better than others.'

'What are these programmes that you like?' Dr Kibe asked.

'One I like is called *Titly*. It's about school education. Let me tell you that it's made very nicely. A lot of drama, songs and stories.'

'Who tells these stories?' Dr Kibe asked.

'There's a Bhil girl called Kamli Didi. And there are two children with her. The programme is very interesting.'

'All right,' I said, 'you liked the programme. But did you learn anything from it?'

He nodded. 'Oh, such a lot! Bhils are illiterate. We've never gone to school. We don't send our children to school either. But now, after watching *Titly*, I take a lot of interest in education. You know what I did? I met a number of villagers and told them that we should start a school. Here in our village.'

'That's very good,' I said.

He frowned. 'But there was a problem. When I talked about opening a school, nobody here would listen to me. They'd say, "What's the point of opening a school and sending our children there. Then who'll do the work at home and in the fields?" So I made them see ISRO programmes on education.'

'Which ones?' Dr Kibe asked.

'There are so many of them! Let me try and remember. There's this *Aao Akash Ko Chhu Le*. Also, *Padhi Lekhi Beti*. There's *Koshish*. Also, *Parchhaiyan*.'

'Then what happened?' I asked.

'After seeing those programmes, they are convinced that our children must go to school. We've now approached the government for giving us a school.'

'That's great,' I said. 'Let me congratulate you for your wonderful effort.'

Mrs Parvati Mohan got us some tea. The tea was nice: piping hot and lightly brewed.

'Do you also watch these ISRO programmes?' I asked Parvati Behn.

Her eyes twinkled. 'Yes, I do. I never miss a programme.'

'Which are the ones that you like?' Dr Kibe asked.

She took some time to think that out. Finally she said, 'I like all the programmes. Like *Abla-Sabla*, *Jara Socho*, *Ped Baba Ki Kahani*, *Bairani Vat*, *Do Kadam Sahi Thor Ki Or*.'

'Which is the one you like most?' Dr Kibe asked.

'*Abla-Sabla*,' she said.

'Why?' I asked.

'Because it's about Bhil women. How they can become strong and powerful. I like it very much.'

'Anything special about the programme?' Dr Kibe asked.

She beamed Dr Kibe a smile. 'Yes, yes. It tells true stories about Bhil women. They even talk to these women in the programme.'

'Any other ISRO programme that you like?' Dr Kibe asked.

'Yes. It's called *Jara Socho*.'

'What's it about?' Dr Kibe questioned.

'About a Bhil girl who travels to all kinds of places in Jhabua. She talks about what she sees. About social issues. And about Bhil women.'

'What did you learn from these two programmes?' I asked.

'Oh, such a lot! They made me think about my life. How useless it has been! That's the time I watched another ISRO programme called *Bairani Vat*.'

'What did the programme show?' I asked.

'It was about Meera Behn from Chulia village. She purchased five goats after taking a loan. Within a period of five months, she had returned the loan. She took another loan of Rs 20,000 and today, she owns a big poultry business. I was very influenced by that programme.'

'So what did you do?' Dr Kibe asked.

'I decided I would take a loan and start a poultry business. I went and talked to a woman in Badi Malpur who had taken a loan and raised chicks. She had made a lot of money. She has paid her loan back and is now going to raise another hundred chicks. She only told me what kind of chicks to rear.'

'Like what?' I asked.

'Oh, all kinds! Even the *Kadaknath* variety.'

'What is this *Kadaknath* variety?' I asked.

She raised an eyebrow. 'Don't you know? It's the chicken with black meat. Very tasty to eat. Well, now I've taken a loan and raised chicks. I'll make a lot of money when I sell them. I'm happy I've done something with my life. Will you come and see my poultry?'

'Of course,' I said.

We went off to see the poultry in Parvati Behn's house. We were joined by Ratan Singh who had just come in. The poultry was in a low-cost kuchha structure that had been added to the house to make room for the chicks. There were a lot of chicks around and they were setting up an awful din. I strained my eyes hard to locate the *Kadaknath* ones, but without success: they all looked the same feathery mass to my urban eyes.

It was time to say goodbye, because Dr Kibe had already reminded me of our next appointment in a village that would take us quite some time to reach.

※※※

Our next destination was Machhlia village. To reach there, we had to go back to Bhabra and take the road to Jhabua. And then on to the Ahmedabad–Indore road.

I dozed off. By the time I woke up with a jolt, we were entering the ghats. The road now curved like a coiled snake: one hairpin bend followed another with relentless regularity, and as our Tata Sumo negotiated the bends hissing and groaning, I could see a cement and glass structure nestling atop a bald hill, looking rather immodest in that stark setting.

'What's that building on top?' I asked the driver. 'Looks like an observation tower. Is it to observe wild life?'

'An observation tower, all right,' he said, his voice gloating. 'But to observe the Bhils, sir. The cops sit in that tower and keep an eye on the road below.'

'Why should the cops do that?' I asked.

'Because this is the famous Machhlia Ghat. The favourite ambushing place of the Bhils!'

We took a turn from the main road to enter Machhlia village. The ISRO TV was kept in the panchayat building. It was there that we met K.K. Jain, the custodian. There were several Bhils there: Dungariya, Sekadiya, Vestiya and Bhikiya. There were two Bhil women too: Ratni and Methalia. We started talking to them.

'Do you watch ISRO programmes?' I asked the Bhil women. 'The ones shown in the evening?'

'Yes, yes.' It was Ratni who answered. She seemed to be the leader of the group. 'The programmes are popular here in Machhlia. We don't miss a single programme.'

'Which are the programmes you like best?' I asked.

'Oh, all of them,' Ratni said. 'But there are some I like more than the others. Like *Ped Baba Ki Kahani*, *Do Kadam Sahi Thor Ki Or* and *Ek Tha Raja*.'

'Why do you like these programmes?' I asked.

'Because they teach us so much,' Ratni explained. She clucked her tongue. 'But why should I be the person yakking all the time? Hey, Methalia, why don't you tell him which ISRO programmes you like?'

Methalia took some time to answer. 'I also like the programmes Ratni said she liked,' she finally said. 'I like *Ek Tha Raja* very much.'

'Oh. Why is that?' I asked her.

'Because it's so nice. You know what it is? There is this king. What's his name, Ratni?'

'Veer Singh,' Ratni said with a flourish. 'People bring their problems to Veer Singh's court. The king turns to his ministers. But they fail to solve the problems. Then a wise man comes in and solves the problems for the king.'

'What kind of problems?' Dr Kibe wanted to know.

Ratni shrugged. 'Oh, all kinds of problems! But there's one particular story we liked very much.'

'What was that?' I asked.

'You see, a Bhil woman comes to Veer Singh's court,' Ratni said. 'She has a complaint to make. That her husband, Dita, is an alcoholic. He has drowned all the money and belongings of the family in alcohol. That he beats her up when he is drunk.'

'Who is this Dita?' Dr Kibe asked.

'Oh, he was a Bhil chief,' Ratni explained. 'Very respected. But then he started drinking.'

'So what did the king do?' I asked.

'The king ordered an inquiry,' Ratni said. 'Dita and his wife were called to the court. The king and the queen explained the bad effects of alcohol. What it did to people, how it made people do bad things that they won't do otherwise.'

'So what did Dita do?' Dr Kibe asked.

'He realized that because of his drinking, he had lost all his money and respect,' Ratni explained. 'He stopped drinking.'

'What did you people do after that?' I asked.

'Something good,' Ratni said. 'All the villagers sat down and talked about it. We decided that drinking was bad. That nobody should drink. You know what we did? We went to Badtal Devi temple in Gujarat. That's the goddess for the Bhils. She gives us whatever we ask for.'

'What did you ask for?' Dr Kibe asked.

'That the goddess should give us strength to give up drinking for ever,' Ratni said. 'All the villagers took an oath before the goddess that they would not drink again.'

'Did the villagers honour that oath?' Dr Kibe asked.

'They did,' Ratni declared. 'They don't drink anymore.'

'What if somebody does?' I asked.

Ratni frowned. She said, 'If somebody does that, he would be fined Rs 11,000. And the money would be credited to the village fund.'

'Have you collected any money so far?' I asked.

Ratni shook her head. 'No, not yet. Nobody in our village drinks now. You know what has happened because of that?'

'Tell me, Ratni,' I said.

She beamed me a smile. 'Something good. Bhils here don't fight with each other any more. And there are no murders. This is the first time in years that there has been no murder in our village!'

'Were there many murders before?' I asked.

Ratni cast a scornful glance at the men present and said, 'Oh, quite a few! Do you know that Machhlia was famous for murders? Every time a Bhil murder took place anywhere in Jhabua district, people said it must have been done by people from Machhlia!'

'How terrible!' I exclaimed.

She nodded. 'Isn't it? But things have changed now. It was all because of alcohol. Another good thing has happened. The Bhils here don't go and stop vehicles on the main road. Do you know that the Bhils of Machhlia were dreaded all over Jhabua district?'

'Is that so?' I asked.

She groaned. 'Yes, of course. Let me tell you that traffic on the road had almost stopped. Nobody wanted to travel after sunset. They used to say that nobody should travel on that road after sunset because the Machhlia gang was at work.'

'Has that changed?' Dr Kibe asked.

'Oh, yes,' she said. 'There has been no incident on this road for the last six months. Why should our Bhils do it? They don't drink any more.'

'That's very creditable,' I said appreciatively.

It was Dungariya who spoke up now. 'These ISRO programmes have taught us all kinds of good things. They've opened our eyes to so many things we didn't understand before.'

'You forgot to tell something,' Vestiya said now. 'You see, we Bhils never went to a doctor before. We always went to a Badwa. The Badwa talked about this witch or that witch! All that silly talk!'

He spat repeatedly on the ground as if to point out how silly the whole thing was.

'Don't you go to the Badwa now?' I asked.

Vestiya shook his head. 'Not any more. After we saw those ISRO programmes, we realized that we had been fools to go to the Badwa. We go to Jhabua hospital when we are sick.'

'Have you learnt anything else from these ISRO shows?' I asked.

Bhikiya Bhil held up his hand. He had not spoken so far. I gestured to him to speak.

He said, 'Dungariya and Vestiya have said correct things only. So has Ratni. These ISRO programmes have taught us a lot. But they forgot to tell you one important thing.'

'What is that?' I asked.

'ISRO programmes have taught us how important education is,' Bhikiya Bhil said haltingly. 'You see, we Bhils are illiterate. We never went to school. But now, we send our children to school. It is only because of the ISRO programmes.'

The driver was at my elbow, telling me that it was time for us to leave. We had to catch a flight at Indore. I bade goodbye to Ratni, Methalia, Dungariya, Sekadiya, Vestiya, Bhikiya and Jain.

We settled down in the Tata Sumo for the long trip back

to Indore. I asked our driver, 'You have been present during all our discussions. What do you think?'

'Incredible!' he exclaimed. 'Who would have ever thought that these wretched Bhils would give up drinking and dacoity? Send their children to school! And go to doctors! If you ask me, ISRO programmes have achieved the impossible.'

As our Tata Sumo turned into the main road, the observation tower nestling atop the hill wove into view.

'That observation tower won't be used now,' I commented.

The driver said, 'Looks like that.'

Was that a hint of remorse I heard in his voice?

2

IN KORAPUT
Finding Drinking Water

If Jhabua is the poorest district in India, Koraput in the state of Orissa comes a very close second. Like in Jhabua, an overwhelming percentage of Koraput's population are from the scheduled castes and scheduled tribes. The district is primarily agricultural. Cultivation of the kharif crop of paddy is the only means of livelihood, but the yield at less than 1.5 tonnes per hectare is so low that it does not give food for more than few months in a year. As if to compound matters, drought is a regular visitor to the district. The result is distress migration and even, absolute hunger. Small wonder, then, that Koraput has often been in the news for hosting starvation deaths.

As if to make up for the poverty, the district offers a rich vista of scenic beauty. With its rolling mountains, undulating meadows, roaring rapids, enchanting waterfalls and terraced valleys leading up to blue hills, Koraput feasts the eye as few other districts do. It is situated on a plateau, with a line of high, verdant hills that boldly mark out its eastern edge. Between the high hills and the low country, there are ranges and ranges of lower foothills, hidden among which are secluded valleys, cut-off from the outside world except for the winding tracks across the passes.

I was travelling in Koraput district to find out how imageries from ISRO satellites have been used to locate sites for drinking water. Travelling with me was Behera, the chief executive of the Orissa Remote Sensing and Application Centre. Behera was in his mid-forties, with an eager and young face and a thick thatch of unruly hair that was just beginning to grey.

It was very pleasant. And cool too, with a nip in the air. We took a turn in the winding road and a tall hill wove into view. Its outlines looked remarkably like the head of an elephant.

'What's that tall hill?' I asked.

Behera peered at the skyline. 'Oh, that one!' he said. 'It's called Hathimali because it resembles an elephant. It's the tallest hill on this side. About 5000 feet high, I would say.'

We negotiated another bend, and there was mist rising from the valley below to greet us.

'Does it happen all the time?' I asked, pointing to the rising mist.

'It's actually the fog,' Behera explained. 'It happens often in the months of November and December.'

'What about the rainy season?'

'The rainy season here is from June to September. During that time, the district is affected by depression from the Bay of Bengal. Particularly in September and October; that brings high wind and heavy rain. In the monsoon months, it rains for days together. The sun doesn't appear for ten days at a time. Low clouds drift into the house and soak everything.'

'What is the average annual rainfall here?' I asked.

'It's about 1520 mm. But the distribution of the rainfall is erratic. It is influenced by the Eastern Ghat group of mountains. The Eastern Ghats run roughly from south-west to north-east. These ghats are like a boundary wall on the eastern side of the district. The monsoon currents rising up the hills cause heavy rain on the hills and their western slopes. By the time the currents have gone over the hills, their water contents have dropped.'

'That means the areas lying east of the hills receive much less rainfall, right?'

'Yes. Those are the rain shadow areas.'

'This drinking water problem in Koraput district,' I asked him, 'is it really acute?'

'Yes, it is,' he said, with a nod of his head. 'It is particularly acute during the summer months. At that time, one sees long lines of women trudging up the hills to fetch drinking water.'

'But why should it be like that?' I asked incredulously. 'Didn't you tell me that the district has an average annual rainfall of about 1520 mm? That's a lot really. With that kind of rainfall, you shouldn't be having problems with drinking water!'

He nodded. 'Yes, the rainfall is adequate. But you see, the problem is with the terrain. People here get their drinking water from the mountain streams and forest springs. In the past, there was enough water going down the streams to serve men and cattle even in the hottest of summer months. That time, the supply of water was almost perennial.'

'And it's not perennial now,' I said. 'Why, what happened?'

'Something to do with the practice of cultivation here. It's called *Podu* cultivation.'

'What does it mean?'

'What the name says. Podu in Oriya means when something is done by burning. People here burn down parts of the forest and clear them so that they can take up cultivation. After a crop or two, they move to another part of the forest to burn it down and cultivate it. That's why it's called shifting cultivation. People in this district have been doing it for centuries.'

'All right,' I said, 'so they do shifting cultivation. But I don't see how it affects the supply of drinking water.'

'You see,' Behera explained, 'in the more densely populated

parts of the district, repeated shifting cultivation over a period of time has choked up the mountain streams. So the streams are no longer perennial as they are so heavily silted up.'

'So the solution is to tap groundwater,' I said.

'Yes,' Behera said. 'You see, in the past, conventional hydrological surveys were done to locate groundwater.'

'What did these surveys do?'

'They collected information on geology and hydrology. And the information was plotted on to topographic maps. But there were problems with such surveys.'

'What kind of problems?'

'Errors used to creep in. Maps were often inaccurate. It was difficult to collect integrated information on many things that control groundwater.'

'How do you set it right?'

'That's why we do it now by satellite remote sensing, because remote sensing is capable of locating areas where groundwater occurs.'

'Where does it occur?'

'In rock formations called aquifers. Aquifers are the ones which help storage and movement of water under the surface of the Earth.'

'Do you see these aquifers from the satellite?'

Behera shook his head. 'No, no. The satellite can't see them directly. But there are certain sub-surface geological elements which form the aquifers. Those elements have surface expressions. A remote sensing satellite can detect them.'

'What are these elements?'

'For example, valley fills. They are narrow valleys formed in the foothills and the adjoining plains. They are filled with sediments over a period of time. Sediments which are loose, unsorted and coarse. Some of these loose materials have different thickness and form good aquifers.'

'Are there other elements?'

'Yes. Like alluvial fans. Fan-like deposits of loose sediments. They generally occur at the base of the hill slope.

Good place for storage and recharge of groundwater. And there are palaeochannels which are basically ancient streams or river beds which are reburied by sediments after the stream has changed its course. Or simply, dried up.'

'But how do they help in the formation of aquifers?'

'Well, they help because they comprise of loose, unconsolidated and unsorted alluvial material. Then, there are the dykes. You see, dykes occur naturally. They are the linear intrusive rocks acting as barriers to water flow.'

'So the satellite reads such information. What do you do with the information?'

'We put the information on to maps. These are the hydro-geomorphological maps. We use them to identify potential areas of groundwater. That's how remote sensing satellites help us to locate groundwater. Well, there's something else. The information given by remote sensing reduces the cost and time involved in locating groundwater.'

'How is that?'

'Let me explain. Because of the information provided by remote sensing, we concentrate our efforts in areas where there is greater potential. And rule out the others. That's how we reduce the cost and time involved in field exploration.'

'Tell me,' I asked, 'Koraput district, what kind of groundwater potential does it have?'

'Koraput is hard-rock terrain. Much of the rainfall here simply runs off the surface. So the recharge is through joints and fractures. There are valley fills in which groundwater is available at shallow depth but they are not dependable sources since they dry up in summer. Also, they cannot sustain continuous withdrawal.'

'Have you suggested drinking water sites in Koraput district?'

Behera nodded. 'Yes, in many places. Right now, we have suggested sites for tapping groundwater in three villages. Electrical resistivity survey has also been conducted for the sites suggested by us in these villages. The survey is done in the site to confirm the location. That gives the thickness of different layers of the soil and weathered materials.'

'When are we going to see these villages?'
'Right after lunch,' Behera said.

※※※

The first village on our list was Boipariguda. It was on the state highway connecting Jeypore and Malkangiri.

'Tell me about Boipariguda,' I said, 'How many sites have you selected?'

'Three sites,' Behera explained. 'The first point is near the college. The second one is in Haldimajhiguda. The third one is in Gondagurha.'

'Difficult names, aren't they?' I said. 'Real tongue-twisters. How do you ever remember them?'

Behera flashed me a smile. 'Oh, I've been to this place so many times!'

'These sites, any particular reason why you chose them?'

'All the three sites are in fracture valleys controlled by lineaments. We've also done the geophysical sounding for all the three sites. They are suitable for dug-cum-bore wells.'

We reached Boipariguda and went to the panchayat office. We were expected. It looked as if all the members of the panchayat were waiting to talk to us. I was introduced to Ramachandra Murjia, the sarpanch of Boipariguda panchayat. He wore a small cloth tied around the waist. He was barefooted. But some members of the panchayat were dressed in a dhoti and kurta while others were dressed in kurta and pyjama. Obviously, modernity was setting in.

Ramachandra Murjia made a short speech welcoming us to Boipariguda.

'What kind of communities do you have here in Boipariguda?' I asked.

'Mainly tribals,' Ramachandra Murjia said. 'We have Bhumias, Parjas, Dhurva and Kandhas.'

One of the panchayat members pointed out, 'You forgot Gaduas, didn't you?'

Murjia nodded. 'Yes, yes, we also have Gaduas in Boipariguda.'

'Which are the major communities here?' I asked.

'The Bhumias and the Parjas,' he said.

'What is the main livelihood for the people?' I asked.

'Most of us have land,' Ramachandra Murjia said. 'But very small holdings. We grow kharif paddy in our land. But the yield is very small. And in most years, there is no crop at all.'

'Why is that?' I asked.

Ramachandra Murjia made a face. 'Because there is drought. Even in times of good rainfall, we can take only one crop. And the paddy we get gives us food only for a few months in the year.'

'So what do you do for food for the rest of the year?' I asked.

'People go to outside places in search of work. They go away for six months. Or sometimes even for longer periods of time.'

'Do the menfolk go alone or take their families when they go out?' I asked.

'Most of the times, they go alone. They leave the womenfolk here at home as well as old people and children.'

'It must be hard for the womenfolk to take all the responsibility of the household single-handed, isn't it?' I suggested.

Ramachandra Murjia grimaced. 'Yes, it is hard on them. In addition to their work, they have to do the work of the menfolk also. And it is particularly difficult for them to fetch drinking water from long distances.'

'Where do the women get the water from?' I asked.

'They used to get them from the mountain streams. But the streams dry up. That means the women have to go deep into the forest to get water from the forest springs.'

'What about the summer season?' I asked.

He made a face. 'That's a particularly bad season for us. We won't have water to drink that time. You have to go about 4–5 kilometres to get a pot of water.'

'So what do you do?' I asked.

'We asked the BDO to do something about it,' he said. 'We passed many resolutions in the panchayat and sent them to the block office.'

'Did anything happen?' I asked.

'Yes. Finally they gave us a scheme, they gave us handpumps.'

'Just plain handpumps? But that doesn't make any sense,' I turned to Behera.

Behera cut in to explain, 'In that scheme, drinking water is provided through handpumps fitted in bore wells.'

'Then, what's your problem?' I asked Ramachandra Murjia. 'That's a good scheme!'

He shook his head. 'It's no good. We've the same problem during the entire summer season. There is no water in the pumps. So our women have to go long distances inside the forest to get water.'

'What's the problem?' I asked Behera. 'Why don't they have water in the pumps in summer?'

Behera explained, 'That's because the bore wells were drilled on wrong sites.'

'What about the sites you have proposed?' I asked Behera.

'They are good sites,' he said. 'You see, groundwater prospect is always site specific. We've selected these sites on a scientific basis. The groundwater prospects are good. And the most important thing is that since the groundwater exploitation is not high here, there shouldn't be any problem about availability of water.'

I turned to Ramachandra Murjia. 'Did you hear what Behera said? That, you'll get water in the sites he has indicated. And plenty of water, too. This time, there'll be water for you even in summer.'

'You know something?' Murjia said with a laugh. 'Behera Babu is our hero. Because here, in Koraput, they have found water only in sites selected by him. We'll certainly find water now. But it's our womenfolk who would be the happiest if that happens. No more travelling long distances to fetch water!'

It was time for us to leave. But before that, I wanted to have a look at the sites myself. So we went to all the three sites that Behera had indicated. To my inexperienced eye, they all looked the same.

※※※

Our next destination was Dashmantpur. Back in the jeep, I asked Behera, 'Now tell me about Dashmantpur. How many sites have you suggested?'

'Three,' Behera said. 'There is this valley to the north of Dashmantpur. We've suggested all the three sites in the valley.'

'Are they good sites?'

Behera nodded. 'Yes. We've suggested the sites at the intersection point of the lineaments where chances of finding groundwater are very good.'

We arrived at Dashmantpur, and went straight to the panchayat office. Subarna Mani Huika, the sarpanch, was waiting for us. Her hair was loosely tied in a knot, with the farthest ends tucked inside. She wore a number of silver bangles on her wrists and a single armlet on the forearms; I could see that the bangles and armlets were rather heavy. She was barefooted.

After the introduction and other formalities were over, I started speaking with Mrs Huika about the village.

'What are the major communities here?' I asked.

'We're all tribals here,' she explained. 'From the Kandha, Parja and Kotia communities.'

'What's the livelihood of the people here?'

'Mainly agriculture. We've small parcels of land converted from forest. The land doesn't produce much.'

'What kinds of crop do you grow?'

'Kharif crop. We take one crop a year and that too, if the rain is sufficient. Although usually there is drought.'

'Does your land give you food for the entire year?' I asked.

She made a clucking sound. 'No, it doesn't. The food lasts for only four months. Not more than that.'

'So what do you do for the rest of the year?'

'People migrate to other places in search of work. Or, they go and pick tamarinds.'

'Tamarinds? Is that so?' I asked. 'What's the season for picking tamarinds?'

'Mostly during summer months: February, March and April. But we've a big problem there. The time to pick tamarinds is from morning till evening. The collection continues till sunset.'

'But what's the problem?'

'You see, we've to take the collection to the agents who are located in the towns. By the time we reach the agents, it is evening. And let me tell you, the tamarind agents in Koraput are the biggest crooks!'

'Are they?'

She squinted, 'Yes, yes. They know that tamarind would come to them in the evenings only, so they bring down the price. They know fully well that the tribal people wouldn't like to go back home with their collection. Even if the price is not right.'

A classical case of exploitation, I thought.

'So what does the tribal do?' I asked.

'What choice does the poor tribal have?' She clucked her tongue. 'If he doesn't accept the price that the agent offers, it means his taking back the whole collection and losing next day's collection also. So he sells the tamarind at the price quoted by the agent.'

'How exploitative!' I pointed out. 'How much does he lose?'

She frowned. 'Quite a lot really! Let me give you an example. The price that the government has notified is Rs 4. That's the minimum price to be paid to us but what the agent gives never exceeds two!'

'That's cheating, pure and simple!' I cried, outraged.

She nodded. 'Yes, it is. But there's even a bigger problem. You see, the whole family goes for tamarind collection. The problem is that the women also have to go to the forest and fetch drinking water.'

'Are you saying that the women can't go and fetch drinking water because tamarind has to be collected?'

'Yes. You see, tamarind is collected in February, March and April. That's also the time when the water sources in the villages dry up. So the women have to travel deep into the forest to fetch water.'

'Do the women get enough water from the forest?'

She shook her head. 'No, they don't. You see, they used to get water from the mountain streams. But they are all silted up. So in summer they are dry. The problem of drinking water is really very acute in the summer months.'

'So what did you do about it?'

'We went and represented the matter to the block development officer. A piped water scheme has been sanctioned. But the main thing is that the wells should be drilled where water can be found in plenty, water that can last us for the entire summer. That's why we depend on Behera Babu so much!'

'Behera was telling me that he has already selected three sites,' I said. 'And all of them in the valley.'

She looked at Behera with reverence as if he was some kind of divinity.

'Do you know what they say about Behera Babu here in Koraput?' she said brightly. 'That, it is only Behera Babu who can tell you where to drill for wells and find water. I know he has suggested sites. The drilling agency will be here any day. That means our problem of drinking water will be over very soon.'

'You have to thank Behera for that,' I suggested.

She smiled. 'Of course, Mr Behera is a big hero of our district!'

※※※

Our next destination was Mathalput village. Back in the jeep, I asked Behera, 'Now, tell me about Mathalput. How many sites have you suggested?'

'Two,' he explained. 'You see, there is a valley near this village. But it is narrow and elongated with low lineament density. But it is the only location that has good groundwater prospect.'

'What do you do, then?' I asked.

Behera's eyes narrowed a fraction. 'There is an added problem. Mathalput village lies on the contact zone between charnokite and khondalite. So it's a faulted zone. There is the possibility that we may encounter thick clay underneath.'

'So what have you suggested?'

'We've not recommended the drilling of a bore well. We've suggested only a dug well. But I'm confident that a dug well, with a large diameter, will yield sufficient water for the village.'

When we reached Mathalput village, the first thing Behera did was to take me to the sites he had suggested for locating the dug well. After visiting the site, we went to the panchayat office. Mrs S. Jani, the sarpanch, was waiting for us. She wore a cotton sari and her hair was tied in a knot. Like Mrs Huika, the sarpanch of Dashmantpur, she wore heavy silver bangles and armlets, but she also wore silver anklets on her legs. She had silver beads on a thread around her neck.

Greetings over, I struck up a conversation with Mrs Jani.

Asking her the same question I had asked the other sarpanches, I gathered that Mathalput too was mainly populated by tribal communities and people belonging to the scheduled castes. The tribals were mainly Parjas, Kotias and Kandhas from the scheduled tribes and the Tombos from the scheduled castes. Agriculture was the mainstay of the people here as well. And as elsewhere in the district, it was inadequate to meet their needs; migration in search of work was rampant, but only after the kharif crop was harvested.

'Is that the only crop you grow?' I asked, about their kharif crop.

'Yes. That is because rainfall is sufficient for growing only one crop. There's no irrigation here. All our lands are on gradients. So the quality of the land is not good. And the yield of the kharif paddy is very low.'

'What about the womenfolk?' I asked. 'Do they migrate along with their men?'

She shook her head. 'No, they stay back here. To look after the old people and children. Also, to look after the household and the cattle.'

'But how do they support themselves?'

'Mainly through forest-based occupations,' she explained. 'They gather firewood, fodder and small timber.'

'Does that give enough to make a living?'

She frowned. 'No, it doesn't. Not now. That's because all the trees have been cut down and very little of the forest survives.'

'Is that so?' I asked.

Her voice rose. 'Yes, yes. At least, the forest near our villages has disappeared. There is some forest left and that is quite far away from the village. So the women have to trek very deep into the forest. And all that they get after that long trek is some dry leaves, twigs, branches, tree roots, shrubs and weeds.'

'Terrible, isn't it?' I suggested.

She nodded. 'Yes. That's the reason why most of our women have given up going deep into the forests in search of firewood, fodder and small timber. They stay at home and do other things instead.'

'Like what?' I asked.

'They make plates out of leaves, or roll bidis. Or, make brooms and mats. But they have a problem with marketing what they make.'

'What's the problem?'

'You see, with the men away from the villages working, the women have to go and market what they make. That takes a lot of time. The question is, if they go to the markets, who'll fetch the water for drinking?'

'Where do they have to fetch it from?'

'You see, the sources near the village dry up during the summer. Even the handpumps don't work. So the women have to cover long distances every day to the mountain streams and forest springs.'

'That takes a lot of time, doesn't it?'

'It certainly does,' she said, frowning. 'Going there, filling your pots and coming back. Particularly, the coming back takes a long time. Since so many pots are balanced on one's head, it takes time.'

'It can be tough also,' I suggested. 'So what are you planning to do about it?'

'We've asked for bore wells to be drilled. The government has agreed. The problem now is to drill bore wells in places where there is a permanent supply of water. You see, we have had bore wells before. But all of them go dry during summer.'

'Hasn't Behera told you where to drill for water?'

She gave Behera a grateful look. 'Yes, he has. We all regard him as our saviour. He has given us those maps and put a cross on the place where the drilling has to be done. He has also given it to those engineering fellows. We are only waiting for them!'

Behera intervened to say, 'You see, we haven't recommended a bore well in this village. I've told the sarpanch that. We've recommended a dug well.'

The sarpanch nodded. 'Yes, he has already told me that during his last visit to our village. The thing is that I always confuse between a bore well and a dug well. How stupid of me!'

Behera smiled at her. 'Most people make that mistake. But let me tell you something, the diameter of the dug well has to be big, so that there is enough drinking water throughout the year. Even during the summer months.'

'God will bless you, Behera Babu!' the sarpanch said in a charged voice. 'Drinking water is the biggest problem for this village now. If a solution can be found, the villagers will be grateful to you forever.'

I teased Behera. 'See, what great faith these villagers have in you! Hope, you don't go wrong.'

The sarpanch shook her head. 'No, no. That'll never happen! You go to any village in Koraput district. They'll tell you that they've got water in places which Behera Babu has selected.'

Dusk was settling in. We could hear the cattle coming home with their bells ringing. It was time for us to take leave of Mathalput and the charming Mrs Jani and return to Koraput for the night's stay.

※※※

On our way back to Koraput, I asked Behera, 'Has ISRO done this kind of groundwater search for other parts of the country as well?'

Behera nodded. 'Yes. ISRO started preparing maps for all 447 districts in the country in 1987. That was for the National Drinking Water Technology Mission. That took five years. The work was completed in 1992. These maps formed a database for the entire country.'

'What happened to these maps?'

'They were useful. Based on these maps, drinking water wells were drilled all over the country. But there were some habitations that were not covered at that time. These are known as the problem habitations.'

'How many of them are there?'

'About 4.3 lakhs. The plan now is to provide drinking water to all these habitations. The project is called Rajiv Gandhi National Drinking Water Mission.'

'Has the work started?'

'Yes. We hope to complete the work soon.'

'Tell me something,' I asked. 'Do you always succeed? What I mean to ask is, do all the sites selected by ISRO yield water?'

Behera shook his head. 'No. Sometimes we go wrong. Let me give you some statistics. The success rate is about 90 per cent. But this is as compared to the 45 per cent achieved using purely conventional methods.'

'That's a big effort.' I complimented.

Behera smiled thinly at my praise. These ISRO guys are so modest, I thought.

3

IN CHAMARAJANAGAR
Educating Children

Chamarajanagar is jinxed. The jinx holds that if a person in high position visits Chamarajanagar, he will lose his position in a matter of days. The story is still told of how a chief minister of Karnataka was out of office a couple of days after his visit to Chamarajanagar. Small wonder, then, that no other chief minister has set foot in Chamarajanagar ever since. Even other ministers have stayed away, fearing the jinx. Needless to point out then that nobody of note ever visits Chamarajanagar.

There was an exception, though. Veerapan, the bandit, did visit Chamarajanagar, and that too, rather regularly. He gallivanted in the district at will in spite of the hordes of policemen hunting him. He kidnapped people with impunity and merrily killed those he did not particularly like; he even hung their disembodied heads from treetops as if in a macabre statutory warning.

With the bandit preying upon the district, life in Chamarajanagar was in tatters. Fear stalked the district like a predator. Quarrying for granite, which is the only commercial activity of note in Chamarajanagar district, stopped. Work in government offices came to a standstill. Patients refused to go to hospitals. Teachers played truant and students who had been brought up on a staple diet of horror stories about Veerappan chopping off people's heads, stayed away from school.

It is not a surprise, then, that the educational statistics of Chamarajanagar district makes for pathetic reading. The literacy rate at 50.87 per cent is the lowest for southern Karnataka and second lowest in the state; it is below the all-India rate and the literacy rate for Karnataka. All in all, the education scenario in Chamarajanagar is pretty depressing.

※※※

That is why it is significant that ISRO took up its maiden programme of distance education for primary schools in Chamarajanagar. I was travelling to Chamarajanagar to see for myself how ISRO's programme was doing. Travelling with me was Shri Bhaskaranarayana (Bhaskar to friends) who is the director of Satellite Communication in ISRO. Bhaskar's official designation sounds impressive, but it does very little justice to his talents. Bhaskar is an ace designer: give him any two places in the world and an antenna; he will find a way of connecting them. But for an engineer who makes communication so utterly seamless, Bhaskar speaks softly. So softly that one has to crane one's neck to hear him.

We started on the long journey and I was happy that Bhaskar was with me. I could use the opportunity to learn about the education satellite that carries the distance education programme for Chamarajanagar.

'Tell me,' I asked Bhaskar, 'why have a satellite only for education? No other country seems to have it!'

'That's true,' Bhaskar conceded. 'No other space-faring country has an exclusive satellite dedicated to education. Even for ISRO, this is our first thematic communication satellite. We call it EDUSAT.'

'What's the general idea? Why EDUSAT?'

'The idea is to share the scarce resources available for

teaching,' Bhaskar explained. 'Creating interactive classrooms by bridging distances. Chamarajanagar is very special to us. That's why it is the very first network that ISRO has established on EDUSAT.'

'How many students can you cover through this network at any time?'

'About 50,000 primary school students.'

'That's impressive,' I said. 'What other networks are you planning on EDUSAT?'

'Quite a few,' Bhaskar pointed out. 'Those will be used for higher education and teachers' training. But they will be interactive. For Chamarajanagar network, we've provided receive-only terminals. They are not interactive as we think that should be adequate for meeting the requirements of primary school students.'

'Fair enough,' I said. 'Tell me about the Chamarajanagar project.'

'You see, ISRO has taken up this project jointly with the government of Karnataka. ISRO has set up 885 terminals for primary schools in Chamarajanagar district.'

'Who looks after these systems in the primary school?'

'It's the local people who maintain the sets. We feel the local custodians can easily maintain the systems.'

'But how do you make sure that the TV set is not misused? That the teachers don't tune in to other channels and have fun?'

Bhaskar beamed me a smile of assurance. 'Well, our TV is so designed that it is tuned to only receive the EDUSAT-based programmes.'

'That's a matter of great relief! Have you covered all the primary schools in Chamarajanagar?'

'Most parts of the district,' Bhaskar pointed out. 'We have been selective. We have selected predominantly tribal areas.'

'What is the programme intended for?'

'Essentially to transmit programmes that supplements curriculum-based teaching in schools. The transmission of the programmes is so synchronized that it corresponds with the school timetable for covering the syllabus.'

'That's nice,' I said. 'But who prepares the contents for these programmes?'

'The government of Karnataka. You know, it's essentially how the contents are prepared that finally decides how successful the programme will be; the content needs to be interesting. They should be such that they sustain the interest of the young audience over a period of time.'

'Are the local teachers here in Chamarajanagar involved in the propagation of these programmes?'

'Yes, they are. The local teacher acts as the facilitator.'

'In what way?'

'They help the children to understand what is being transmitted. And also clear doubts if there are any.'

※※※

We arrived at the Government Primary School in Palya village. The building that housed the school was the usual rectangular, nondescript structure built in a style favoured by government architects about three decades back. We went in. We were expected. We were introduced to the teachers—Shri B. Shivaiah, the head master, Shri Govindaraju and Shri Siddaiah, the teacher-in-charge of the EDUSAT programme.

I looked around the room. It was not particularly tidy. The furniture was crumbling. Most desks wobbled on three legs instead of the mandatory four. The blackboard was no longer black, but an indeterminate grey with the remains of yesterday's lessons scribbled in shapeless squiggles. The elegant cabinet that housed the ISRO TV and accessories, looked rather immodest in that untidy setting, much like the circular dish antenna and the rectangular solar panels outside gleaming in the afternoon sunshine.

I could see about sixty students with shining faces and stained dresses, sitting around the room. They sat there in uneasy intimacy with their eyes riveted on the cabinet, and occasionally nudging each other in impatient gestures. There was an expectant air about the room. It looked as if all they wanted was the cabinet to open and the TV to come to life.

Action started at about 1.50 p.m. Siddaiah, the teacher-in-charge, stood up as if to address the gathering. He was a tall, hefty and bespectacled man, with a strident voice befitting a village schoolmaster who was required to discipline a motley bunch of unruly students.

He started explaining about the TV programme that was to follow. He spoke in a manner the children could understand. As if in affirmation, the children nodded their heads vigorously to indicate that they understood.

I turned to Bhaskar. 'I think Siddaiah did a good job,' I pointed out. 'He managed to explain to the students about the programme. Tell me, is that a requirement?'

Bhaskar nodded. 'It's called the pre-broadcast interaction. It's a part of the project design. When the teachers-in-charge are trained, they are told how to structure the pre-broadcast interaction.'

Siddaiah finished his pre-broadcast interaction and there descended on that room a hushed, suspenseful silence. A silence so intense that I could now hear large raindrops battering the glass planes of the windows. So it had started raining heavily, I thought.

As if on cue, the electricity went off. But there was nothing unusual about that, because in Karnataka's villages it is almost a law for electricity to play truant at the slightest hint of rain. Will it affect the TV transmission, though?

'Electricity's gone! What will happen now?' I asked Bhaskar. 'Will it affect the transmission?'

Bhaskar beamed me a smile of assurance. 'No, it won't. The system will now operate on the solar panel.'

'Oh, does it?' I exclaimed in a relieved voice. 'I didn't know that!'

'It does. We did think of this contingency when designing the system,' Bhaskar explained. 'That's why we put in the solar panel. Maybe it makes the system slightly more expensive, but the point is that it works in times like this.'

As if to confirm what Bhaskar said, the transmission did start in spite of the electricity playing truant. And there was a burst of loud applause from the students. The picture that appeared on the TV was crisp and clear. The audio was good, too. Obviously, the rain pouring outside had not affected the quality of transmission.

The programme was about the need to know the history of one's family and the steps to be taken to record one's family history. The programme was done quite well. I could see that the students were watching the programme with a great deal of concentration. But was there a faint frown on most faces? As if they couldn't quite understand what was going on. I made a mental note to check that with the students once the programme was over.

The main programme finished and Siddaiah stood up. Was he going to deliver another lecture? Questions now flashed on the screen, with sufficient time for the students to answer them. The questions related to the subject of the transmission and I noted that most students were able to answer them. Siddaiah now asked the students to get their doubts clarified. There were none.

'What was that?' I asked Bhaskar.

'Oh, that was the post-broadcast interaction,' Bhaskar explained. 'That's part of the scheme. It has to be conducted by the teacher-in-charge after each transmission.'

Bhaskar and I walked across to a boy and girl who were discussing the programme in a corner of the room.

'What's your name?' I asked the girl.

'My name is Kanyashree,' she said shyly.

Kanyashree was tall for her age, with a face that assumed a startled expression every time a question was addressed to her. Like a doe, I thought. She was dressed in a skirt and blouse that had seen better days.

'In which class do you study?' Bhaskar asked.

'Class Five,' she said laconically.

'Why were you frowning?' I asked her. 'Was there something wrong with the programme?'

'Did I?' she stammered out, looking flustered.

I could see that she was thinking my question over in her mind. After a while, her face brightened.

'Oh, that!' she said with a giggle. 'I had some problem with understanding the language.'

'Why should that be?' I asked, perplexed. 'The programme was in Kannada!'

She giggled again. 'Yes, it was in Kannada. But it's not the kind of Kannada we speak at home. It was different.'

Bhaskar intervened to say, 'I know what it is. The person in the programme used the Kannada dialect spoken in the northern part of Karnataka.'

'But did you understand what was said in the programme?' I asked Kanyashree.

She nodded. 'Yes, I did. With some difficulty. I could broadly understand what was being said.'

'Is it like that on other days of the transmission?' I asked.

She shook her head. 'No, no. The programmes on other days were in the kind of Kannada we speak at home.'

'Did you like the programmes on the other days?' Bhaskar asked.

'Yes. Very much.'

'But the programmes are on subjects from your text book only!' I pointed out. 'Don't the books tell you the same thing?'

'Yes, but when they come on the TV in these programmes, they help me to understand my lessons much better. Also . . .'

Her voice trailed off.

'Tell me,' I coaxed her.

She found her voice at last. 'Sometimes, these programmes say new things. They also say them in a different way.'

'In an interesting way,' I suggested.

She nodded vigorously. 'Yes, yes. That's what it is. It's much more interesting.'

She paused for a while. I could see that she wanted to tell us something more. Obviously we had won her confidence.

She did not have that startled look when we asked her questions.

'Do you know,' she finally said with a blush, 'that I look forward the entire day for the programme to come on the TV?'

'Do you have a TV at home?' Bhaskar asked.

She shook her head. 'No,' she said.

'Is that why you find the programme here so interesting?' I asked.

She thought for a while, then said, 'Maybe. But I also like it because the programme helps me to understand my lessons better.'

It was a long interview for a small girl. The effort had obviously tired her. I told her she could go. She ran off, looking relieved.

I turned to the boy. 'What's your name?' I asked.

'Praveen,' he said.

Praveen looked clever. He was dressed in khaki half-pants and a shirt that was white once but now looked like a patchwork quilt with a number of yellow food stains and purple patches that suggested too many messy trysts with the jamun tree in the school compound.

'Which is your class?' Bhaskar asked.

'Class Five,' he said.

'Do you like these TV programmes in school?' Bhaskar asked him.

He beamed Bhaskar a smile. 'Very much, sir. These programmes are the only interesting thing in the school here.'

He looked around furtively. Was it to make sure that nobody was eavesdropping on our conversation?

He lowered his voice and said, 'The teachers here are not of the same standard as the people in the TV programmes.'

'Why do you say that?' I asked.

His voice dropped to almost a whisper. 'The teachers here, sir, always go to the blackboard and write something there. We have to copy whatever is written on the blackboard. Most of the times, I don't even understand what I am copying. The teachers don't explain.'

'Is it different in the TV programme that you see here?' Bhaskar asked.

He nodded. 'Yes. The TV programmes help me to understand the lessons in my books much better.'

'Are they interesting?' I asked.

His face brightened. 'Yes, yes. Very interesting. Because they tell everything in the programme like in a story, I can understand everything.'

'What about the picture on the TV?' Bhaskar asked. 'Does it come nicely?'

He beamed. 'Yes, yes. Just like the TV we have at home. You know, we've Udaya TV. The picture here is just like that.'

It was my turn to be perplexed. 'What's this Udaya TV?' I asked him.

'Don't you know?' Praveen asked in surprise. 'Don't you have Udaya TV in your home too?'

Bhaskar chipped in to explain, 'What Praveen means by Udaya TV is actually cable TV. You see, in these parts of Karnataka, they mostly watch Udaya channel on the cable TV.'

I told Praveen, 'Kanyashree was telling us that she waits the whole day for this programme to come. Is it like that for you?'

Praveen made a deprecatory gesture as if Kanyashree's opinion did not amount to much.

'Did she tell you that?' he said. 'Yes, it's like that for me too. Other students here also wait for the TV programme to come. Can I go now, sir?'

'Yes, of course,' I said. 'Thank you very much.'

He ran off. He was half way down the room when he retraced his steps and came towards us.

'Just one thing, sir,' he said with a conspiratorial air. 'Please don't tell my teachers here about what I told you.'

He ran back. We walked across to the corner where the teachers were standing.

Siddaiah asked me, 'We saw you talking to the students. Was it about today's transmission?'

I nodded. 'Yes,' I said.

'What did they say?' he asked.

It was Bhaskar who replied. 'They said they found the programmes very interesting; also, that the programmes helped them to understand their lessons better.'

'That's what we also think,' he said.

'Let me ask you something in general,' I said. 'What do you think of the ISRO's EDUSAT project?'

Siddaiah smiled at me. 'It's a fantastic project. ISRO must be congratulated for sending up a satellite only for education.'

The head master chipped in. 'We all think so. It is a great thing that ISRO takes so much interest in education that it has made a satellite for education only.'

Govindaraju spoke now. 'That too in primary education. Nobody really bothers about primary education. A relative who lives in Bangalore was telling me that they use this kind of distance education for management classes. But to think of ISRO doing it for primary education and connecting places like ours! We are really the back of beyond!'

That was indeed a long speech, I thought. May not be by a teacher's standard, though.

'But are these transmissions useful?' Bhaskar asked.

Siddaiah answered, 'They are useful to the students. The students can understand their lessons better after they've viewed the transmissions. You should see the eagerness with which they wait for these programmes!'

'What about the contents of the programmes?' I asked. 'How good are they?'

'The contents are of top quality,' Siddaiah explained. 'I would say that the contents have been prepared with a lot of imagination and creativity. They use the story-telling mode and the students seem to like that.'

Govindaraju intervened, 'The students like it because it comes on the TV. You see, most of these students here come from poor families and they don't have TV sets at home.'

Siddaiah protested, 'It's not merely the TV, it's also the quality of the contents.'

'How is the quality of the audio and video?' Bhaskar asked.

It was the head master who commented now, 'Excellent. Just like it is in the commercial channels.'

I turned to Siddaiah. 'Have you been given training for this EDUSAT programme?' I asked.

He nodded. 'Yes, I have undergone one spell of training. Let me tell you that it was very useful; with what I learnt during the training programme, I can conduct the programmes here in the school very well.'

'Now, coming back to your pre- and post-broadcast interactions with the students,' I asked Siddaiah, 'are they useful?'

'Yes,' he explained. 'They are useful. When I give the pre-broadcast explanation, I prepare the students for the programme to come. That way, they are in a better position to appreciate the contents of the programme.'

'What about the post-broadcast interaction?' Bhaskar asked.

'That also helps a great deal,' Siddaiah said. 'There are questions that flash on the screen. I help them along to answer them. Then, there are doubts that they have. I address those doubts.'

'Tell me one more thing,' I asked Siddaiah. 'Do you get programme schedules well ahead of the programmes?'

Siddaiah nodded. 'Yes. They circulate the programme schedules to all the schools well in advance. The schedule is very detailed. That helps in preparing for the pre-broadcast interaction with the students.'

'Let me ask you something in general,' I said. 'You people say that these programmes help the students to understand their lessons better. Doesn't that threaten your status as teachers?'

'No, it doesn't,' Siddaiah said. 'Why should it? In fact, to tell you the truth, we feel empowered by these programmes.'

The head master spoke now. 'What Siddaiah is telling you is correct. All of us are very enthusiastic about the programme. We feel that it has given the entire teaching community a lot of power.'

'Any suggestions about possible improvements to the programme?' Bhaskar asked.

'Only one thing,' Siddaiah said. 'About the teacher training programme that they conduct under the EDUSAT scheme. That was useful. I think they should conduct them regularly and on a continuing basis.'

The headmaster chipped in. 'Why don't you talk to some parents and find out what they think of the programmes? In any case, you must have a cup of tea with us before you leave.'

The headmaster led us to two parents and introduced them to us. One was Siddaraju and the other, Nagaraja Naik. Siddaraju looked affluent. He was dressed in immaculate white. His dhoti was folded neatly just above his knees and he wore a white shirt, with a towel thrown on it. He had an air of assurance about him.

'Do you also watch these EDUSAT programmes?' I asked Siddaraju.

He laughed, showing a set of even, white teeth. 'No, no. I just came here to collect my son. You see, it's raining heavily and he had forgotten to bring his umbrella with him.'

'Oh, I see,' I said. 'But do you know about these programmes.'

He nodded. 'Yes, my son tells me all about them. He is very excited about these programmes.'

'Why is he excited about them?' Bhaskar asked.

'He says he finds them very interesting. Also that, after he sees these programmes, he understands his lessons much better. Let me tell you something else. On the days the TV transmission is there, he gets ready before time and waits to go to school. On other days, he is so reluctant that I have to push him hard.'

Nagaraja Naik coughed meaningfully. I looked at him. He looked like a man who had made his living with his hands. Like Siddaraju, he sported a dhoti folded above his knees, a shirt and a towel, but his clothes were crumpled and suggested that they had seen better days. I knew he wanted to say something.

'What do you think?' I asked him.

'What Siddaraju says is correct,' he said. 'My son also studies in this school. After the TV programme has started here, he attends school very regularly. Before that, he used to be absent most days.'

'What happens now?' Bhaskar asked.

'He never misses school now. You see, we belong to the Nayaka community. That is a scheduled tribe. Nobody from our family had ever gone to school. My son is the first one to go to school.'

'Is that so?' I asked.

Nagaraja Naik nodded. 'Yes, yes. He's the first one. Since we are not educated, we're not able to help him with his studies. He does it on his own. Now he says the TV lessons have helped him a great deal.'

'In what way?' Bhaskar asked.

'Because he says, with what he learns from the TV lessons, he can now understand what is written in his book.'

Siddaraju chipped in. 'I forgot to tell you people something. My son says that the way they narrate it in the TV lessons is like telling a story. So it's very interesting. Just like the serials we see at home on Udaya TV.'

He gave Nagaraja Naik a condescending look. Was it to convey that he had Udaya TV at home while Nagaraja Naik did not?

The headmaster approached us to announce that our tea was ready. The tea was served to us in chipped glasses, but the tea was good—hot and creamy. We finally took leave of them.

※※※

It was still raining when we left Palya. Our next destination was the Government Primary School at Saragur. We were supposed to catch the transmission there at 3.30 p.m. Our progress was slow because the road had become slick and slippery with rain, which still continued to pour. I said a silent prayer for the rain to stop so that we could reach Saragur in time for the transmission. Somebody up there was obviously listening: the rain stopped and we reached Saragur well in time.

The school building at Saragur wore a more modern look. It had obviously been built later than the school building in Palya. There was the dish antenna, sitting arrogantly atop the building, with rainwater falling from it in a rhythmic drip-drop as if somebody had put a sprinkler set up there.

The science teacher who was also the teacher-in-charge for the EDUSAT programme was waiting for us. There were about fifty students who sat there, their faces bright in anticipation of the programme to start.

The programme started almost immediately. The science teacher began his pre-broadcast explanation. The programme, the science teacher explained, was about how plants made their own food. He was able to convey the importance and essence of the subject matter of the transmission about to follow.

The transmission started on time. The quality of both audio and video was excellent. The lesson on plants preparing their own food was done rather creatively in the manner of telling a story, the language and style simple enough for the students to understand and still find it interesting. The man presenting the programme at the studio's end was articulate and knowledgeable; it was clear he knew the subject well.

When the transmission on the lesson ended, the students clapped. Obviously, they had liked it. The science teacher started his post-broadcast interaction. Questions flashed on the screen and gave some time for the students to answer them. I could see that the students were able to answer the questions within the time allotted. Finally, the transmission was over, and the science teacher asked for any doubts that remained to be addressed. There were no questions.

It was time for us to interact with the students. I asked the science teacher whether we could speak with some students. He readily agreed and selected four of them—Suresh, Chitra, Madhushir and Ashwini.

The students selected were a motley bunch. Suresh had a very intelligent face and diligent manners. He wore clothes that were almost in tatters and in any case, several sizes too large for his puny body.

'Tell me, Suresh,' I asked. 'Did you like today's programme?'

'Yes, I did,' he said in a clear voice.

'What about the programmes on other days?' Bhaskar asked.

'I liked them too,' he said.

'Now tell me,' I asked him. 'Why do you like these programmes?'

There was no hesitation in his voice as he said, 'Because the teacher who explains the subjects in the TV programme is a very good teacher. He knows the subjects very well. He is able to explain everything.'

That sounded like a very adult answer. 'Is that so?' I asked.

Suresh nodded. 'Yes. You see, in the subjects that have come in the TV programmes here, all my doubts have been cleared. I can answer any question on those subjects.'

'What about your teachers here in this school?' Bhaskar asked.

He didn't even look around to see if the teachers were listening in before he answered.

'They are not that good. Certainly not as good as the teacher who explains the subject on the TV. That teacher is the best on that subject.'

'How are you doing in the class?' I asked.

'Well, sir, I always come first in class. You know what my father says? That I am not getting a proper education in this school. So he's planning to put me in a better school either in Mysore or Bangalore.'

'Is that so?' Bhaskar asked. 'Then why don't you go?'

Suresh made a face. 'Because we are very poor, sir. My father doesn't have enough money to send me to those schools. He's talking of taking a loan. But I know he can't pay back the loan. Where will he find the money from? We're really very poor, sir.'

'So what will happen then, son?' I asked.

'I've told my father not to worry. Things have changed after these TV programmes have started in school. Only the best teachers teach in these TV programmes.'

'You really think so?' Bhaskar asked.

'Yes, sir. It is like I can sit here in this room and listen to the best of teachers on the TV. There is no need for me to go anywhere else.'

There was a certainty in his voice that told me his mind was made.

'Have you told your father?' I asked.

'I've told him, sir. A number of times. But he won't listen to me. He worries for my future.'

I looked at him in admiration. This boy was such a gem. Here he was, a boy from a poor family, studying in a primary school in the deep recesses of Karnataka's tribal belt, and yet, he was so intelligent and spoke like an adult. Much like the clothes he wore, which were several sizes too large for his tiny frame, he was wiser beyond his young years.

Somebody from the other end of the room called out Suresh's name, and he asked me whether he could go.

I smiled at him. 'Of course, you can.'

He ran off. I turned to the girls. Chitra was dressed in a skirt and blouse ensemble that looked quite new. Her hair was well oiled, with two plaits that were tied together by a colourful ribbon. She looked smart and ready to talk. I got the impression that given half a chance, she would talk her head off. By comparison Madhushir and Ashwini looked subdued. The clothes they wore suggested that they had seen better days.

'Did you watch the programme today?' I asked Chitra.

She beamed me a radiant smile. 'Yes, I did.'

'Did you like it?' I asked her.

'Yes, very much. I liked the way they told the story.'

'Is it one of your lessons?'

'Yes, yes,' all the three girls said in unison.

I suppose that's the way the students respond to questions by teachers—assent in a chorus.

I next turned to Ashwini. 'Have you read the lesson in your book? Before today's transmission, I mean.'

Ashwini nodded vigorously. 'Yes, but I hadn't understood much of it.'

'Why?' Bhaskar asked her.

She looked at Bhaskar in some puzzlement. Bhaskar repeated his question a little more elaborately.

Ashwini smiled. 'Because, in the book, they say it in a way I find difficult to understand. The teacher has also taught it. But even then, I didn't understand.'

'But what about the TV programme today?'

It was Chitra who answered. 'It's only after I saw the programme today that I understood how plants prepare their own food.'

'Because they tell it like a story?' I asked.

'Yes, yes,' all three of them chorused in unison.

It was Madhushir's turn now. 'Did you also like the other programmes that came on ISRO TV?'

'Yes, sir. They are all good,' she said, 'they help me to understand my lessons better.'

Chitra cut in, 'It's just like the programmes on Udaya TV we have at home. So interesting!'

'Is that what both of you also think?' I asked Madhushir and Ashwini.

'We don't have Udaya TV at home,' Ashwini said with a deep frown.

'Me, neither,' Madhushir said and looked venomously at Chitra for having brought the topic up.

'What do you people think of the timing of the programmes?' Bhaskar asked. 'Should it be in the morning or in the afternoon?'

Chitra said excitedly, 'it should be in the morning, as soon as we come in. Then, I don't have to wait for the whole of the morning to pass to see the programme.'

Madhushir chipped in, 'What Chitra says is correct. It should be in the morning.' Now, she glared at Chitra and said, 'We're not as lucky as Chitra. We don't have TV at our homes!'

Chitra glared back at Madhushir and said in a raised voice, 'What's wrong with having Udaya TV at home? Even then, don't I wait impatiently for the school programme to start?'

The science teacher approached us now and asked the girls, his voice rather strident, 'What's going on here? What's the ruckus about? Can't you children answer questions without fighting amongst yourselves? Is this the way to behave in front of visitors who have come from such a long distance?'

The girls looked down with penitence writ large on their young faces. I intervened to say, 'They were not fighting. They were merely competing with one another to tell us how eagerly they look forward the whole day to watch the transmission.'

The expression on the science teacher's face softened.

'That's true,' he said in an indulgent voice. 'All the students here wait eagerly for the transmission.'

'Do you sometimes feel threatened by all this enthusiasm of the students for the TV lessons?' I asked the science teacher.

He now addressed the students in a teacherly tone, 'What are you doing here listening to elders discussing serious matters? Off you go!'

The girls scampered back hurriedly to the corner of the room where the other students had gathered and were raising a din.

'Now, about your question,' he said in an even tone. 'No, I don't feel threatened. Why should I? This is a good programme. In fact, I regard this as a programme that helps me to teach my students better. I welcome it.'

'That's nice of you,' I said.

'I suppose you would like to talk to the other teachers?' he suggested.

'We would like that very much,' I said. 'Only if it can be arranged without causing any inconvenience.'

'No problem,' he said. 'I'll request them to come here and meet you.'

He called out to one of the students to go and fetch the other teachers.

'Let me ask you something before the other teachers come,' Bhaskar told the science teacher. 'Do you think this programme has helped the students here?'

The science teacher nodded. 'It certainly has. Greatly, in fact. It has helped them to understand their lessons much better.'

Two other teachers trooped in. One of them announced, 'I'm the physical education teacher here. Why is it that lessons on physical training are not included in the transmission? I think there should be programmes highlighting physical fitness.'

'That's a nice suggestion,' Bhaskar said. 'But does that mean that the existing programmes in the transmission have helped the students?'

The physical education teacher fingered his moustache. 'Yes, it has helped them. You should see with what eagerness they wait for these programmes! What do you say, Mrs Vijayalaxmi?'

Mrs Vijayalaxmi, the other teacher, nodded. 'I agree. The programme has helped the students. That's why there is a need to use the medium for more number of hours and include other subjects as well. There is one more thing. You see, the contents of the English lessons in the TV programmes are not all that useful.'

'Why is that?' I asked.

'That is because English, at least in the beginning, should be taught in Kannada. And also, the contents should be integrated with the curriculum.'

'Any other suggestions?' Bhaskar asked.

It was the science teacher who spoke now. 'Just that some of the good teachers from this area should be chosen to give lectures on some of the topics.'

The teachers suggested that we have a cup of tea with them. We went to a corner of the classroom where I saw that tea had already been brought in and was about to be served. It was served in proper cups, unlike the chipped glasses in Palya.

It was while we were having our tea that we heard the sound of distant thunder, ominous and threatening. I looked through the window. The sky was full of mud-brown clouds heavy with water. We said goodbye to the teachers and took leave of the students. The rain came now in huge drops. We got into the car and settled down for the long journey ahead of us. Soon, the heavens opened and raindrops fell noisily on the car clattering against the metal of the roof.

Amidst all that noise, what Suresh had said kept ringing in my ears: that all the best teachers came to teach him in his little village school through the ISRO TV and that he had no need to go to a better school in a bigger city.

I now realized what the EDUSAT was all about. It brought the best teachers virtually to remote and inaccessible places to teach students who were not privileged enough to be in big cities.

It took a young boy from Saragur to make me understand that.

4

LAKSHADWEEP
Helping Fishermen

Saint Ubaidullah is a revered name in the Lakshadweep Islands. The story is still told of how the saint used to fall asleep ever so often and then, dream. Once while praying at Mecca, the saint fell asleep. He dreamt that the Prophet wanted him to go to a distant place. He left, but after sailing for months, a storm wrecked his ship near the Lakshadweep Islands. Floating on a makeshift raft, he was swept ashore. He fell asleep and dreamt of the Prophet asking him to propagate Islam in the island. He did that. This in turn enraged the headman of the island. The headman and his cronies surrounded the saint to kill him. He beseeched the Almighty and the people were struck blind. It was only after the saint left the island that the people regained their eyesight.

To lose your eyesight can possibly be the worst thing that can happen in the incredibly beautiful Lakshadweep Islands. The thought crossed my mind as the chopper taking us to Kavaratti made its descent. From the air, the island looked like an emerald in the vast expanse of the translucent sea that sparkled and shimmered in varying hues of azure. The island was fringed by snow-white coral sands, and there was a calm lagoon separated from the incoming swells of the outer sea by coral boulders. It was stunning.

The chopper landed, and we waited for the blades to stop rotating. Through the tiny windows, I could see the large, portly figure of Dr Koya, the director of Science and Technology, hurrying towards the aircraft with an exasperated expression. Why the exasperation? Were we unwelcome guests?

Dr Shailesh Nayak and I were visiting Kavaratti to learn how ISRO had helped the fishermen of the island. Dr Nayak, a brilliant application scientist, had done more than anybody else in ISRO to devise and put in place useful initiatives for fishermen to increase their catches. He looked like a successful Gujarati businessman, complete with an intense expression and amiable manners, but that only proved how deceptive appearances could be.

The blades stopped rotating and we got down the narrow stairs. At the foot of the stairs, there was Dr Koya to greet us.

'The blasted chopper!' he said, his voice panting. 'It can never keep time! Never! Today, it landed much before its time. My blessed luck! I almost broke my back trying to be here in time before the chopper landed.'

Well, that explained his exasperation. As we drove along the only cemented road the island boasted of, the car was full of Dr Koya's panting. I waited for it to subside.

'These islands are incredibly beautiful,' I observed. 'How many of them are there?'

'We have thirty-six islands, twelve atolls, three coral reefs and five submerged coral banks,' Dr Koya explained, still gasping.

'Oh, that many!' I said. 'How big are these islands really?'

'Not particularly big,' Dr Koya said. 'They vary from 0.1 to 4.4 square kilometres. But there's something interesting about the layout of these islands. It is more or less similar. The island is always in the east, the lagoon is in the west and the island tapers down towards the south. This is true of all the islands here.'

'The layout is certainly an interesting coincidence,' I agreed. 'I also saw a large number of lagoons from the air.'

'Did you?' Dr Koya said. His panting had totally subsided and his voice was even now. He continued, 'Yes, there are quite a few of them. Fifteen of them to be exact. The area they cover is considerable too. About 420 square kilometres.'

'That's a big area,' I agreed. 'What do these lagoons do?'

'They are very shallow,' he explained. 'But they support a rich growth of macrophytes. The algae in the lagoons play a big role in the biogeochemical processes of these waters.'

'You mentioned something about the coral reefs, didn't you?' I asked. 'What are they?'

'They are compound atolls with a westerly or centrally located lagoon system. They are rich in species diversity.'

'How wide are these lagoons?' I asked.

'About 1–4 kilometres. But they vary in depth. Small lagoons, like the one we have here in Kavaratti, are filled with sediments with a depth of 1–15 metres.'

'What about the large lagoons?' I asked.

'The large lagoons, like they have in Bangaram, are deeper with depths of 10–25 metres. These deep lagoons harbour coral knolls that are sites of high coral diversity and continuous coral growth.'

Dr Shailesh Nayak, who was quiet all this time, now chipped in. 'These coral reefs are always affected by what is happening to the ocean. Let me give you an example. Even slight, short-term changes in the normal oceanic temperature result in coral bleaching. But long-term changes in the temperature stress the corals significantly.'

'Tell me, Dr Koya.' I asked. 'The marine resources of the islands should be pretty rich, aren't they?'

'Yes,' Dr Koya said. 'The lagoon, sandbanks, open reefs and submerged banks which form part of the archipelago are very rich in marine life and mineral resources. They are spread over an area of about 4200 square kilometres.'

'That huge?' I asked.

'Yes. There's something interesting there. They've extended India's economic zone by about 4,00,000 square kilometres.'

'But what are the kind of things you have there?' I asked.

'We have fish, crabs, corals, sea star, sea cucumber, sea horse, jelly fish, octopus, shells and turtles.'

'Turtles?' I asked. 'What kind of turtles?'

'There are four species of turtles. The most common are the green and hawksbill turtles. We sometimes see Olive Ridleys outside the reef or in the lagoon.'

'What about shells?' I asked.

'Plenty of them. Money cowries are abundant in the shallow waters of the lagoons and in the reef. They are usually picked up from the reef area at low tide by the womenfolk during their spare time.'

'Where's the market for them?' I asked.

'Largely in the mainland. They are used for decoration. Cone shells are also quite common. We also see the clam in the coral boulders exhibiting its brilliantly coloured mantle to camouflage itself from its surroundings.'

Dr Koya was becoming lyrical, I thought. But then, anybody would do that, if called upon to describe the brilliantly coloured mantle of the clam.

'You said something about octopus, didn't you?' I asked.

He nodded. 'Yes, I did. We see them in the crevices of the rock boulders. Incidentally, the octopus is considered a great delicacy by the islanders.'

'What about the fish?' I asked.

'We have tuna, shark, sardine, mackerel and silver belle. But it's mainly tuna. We have the coastal tuna, known as the little teenie. It's landed by surface fishery, using pole and line gear fishing. But the main activity on which the economy of the islands depend is tuna fishing of the deep sea variety. That's the stuff for export, meant for the international market.'

'The sea around the islands is very productive, isn't it?' I asked.

'Yes. The per capita fish landing in the islands is about 198 kilos a year. You see, our islands stand first in the country in per capita availability of fish.'

'Have you estimated the potential fishery resources of the sea around the islands?' I asked.

'Yes, we have. We estimate that the potential landings of the Laccadive seas are of the order of about 90,000 tonnes. But currently, about 30,000 tonnes are being exploited. But tuna is the major fishery for the islands.'

'The chicken of the sea?' I suggested.

He nodded. 'Yes, that's how it's described. But tuna is available in plenty around the Lakshadweep Islands.'

'What kind of tuna?' I asked.

'Mainly skipjack and yellowfin. They contribute about 99 per cent of the tuna caught here during October and May. The fishermen here go largely for tuna fishing. After they catch the tuna, they dry it in the sun after cooking and smoking. The dried tuna is called *Mas*. The *Mas* sells for about Rs 75–90.'

'Don't you have mechanized facilities for cooking, smoking and drying tuna?' I asked.

'Yes, we have, but that's only at Minicoy. They have a canning factory there to process fresh tuna.'

'What about mechanization in fishing?' I asked. 'Have you introduced it in the islands?'

'Yes. Previously fishing here was done with locally made wooden crafts and traditional implements. It was laborious, time-consuming and uneconomical. We've now introduced modern, mechanical fishing.'

'Has the fish catch gone up as a result?' I asked.

'Enormously. Let me give you some numbers. In 1960, we had old country crafts and the total landing of fish was only 600 tonnes. The advent of mechanization took sometime to be completed. By 2001, when mechanization was complete, the fish landing was more than 10,000 tonnes.'

'That's very heartening,' I said. 'On that note, why don't you tell me about the people on the islands?'

'Well, the entire indigenous population of the islands are classified as scheduled tribes. That is because they are socially and economically backward. But something interesting has happened about the population. It has grown considerably in the last 100 years.'

'By what percentage points?' I asked.

'By 337 per cent since 1901. But there's one distinctive thing about the population of the islands. They maybe socially and economically backward, but everyone in the islands goes to school. You see, all of them are literate. In fact, in terms of total literacy, it is the third highest in India, only after Kerala and Mizoram.'

'That's very commendable,' I pointed out.

'But what is worrying is the growth rate of the islands. In terms of the growth rate, it is twenty-ninth in India. Almost at the bottom of the list. That's because there is nothing much in the islands for the people to do gainfully. The main occupation here is fishing, coconut cultivation, and coir twisting. In any case, fishing is the most important occupation and the entire economy of the islands depends on how productive fishery is as an occupation.'

'What you mean to say is that the fish landings have to improve still more substantially for that to happen,' I suggested.

He nodded. 'Exactly. The improvement in fish landings because of mechanization has been a step in the right direction. We are hopeful that the fisheries forecast based on satellite data will boost the fish landings even further.'

We had reached the Inspection Bungalow where we were to spend the night. The bungalow was right on the sea and the view was indeed stunning.

'What's the plan for the day?' I asked Dr Koya.

'I'll be back in the afternoon to take you to my office,' he announced. 'Just to show you how we receive the satellite-based fisheries information and disseminate it.'

'Aren't you taking us to meet some fishermen?' I asked. 'We would like to know how the information has helped them.'

'Yes, that has been planned,' Dr Koya said. 'That's in the evening. I've called some user fishermen to my office; you'll meet them there.'

※※※

Dr Koya left. We had our lunch in the veranda that overlooked the sea. It was a simple meal of rice and a tasty fish curry with a green salad thrown in. The cook stood solicitously by our side as we ate our lunch, if only to make sure that we ate heartily and well.

'What's your name,' I asked him.

'Fabrikage,' he said in a singsong that wavered like the sea by our side. 'I've worked as the matey here for the last twenty-five years.'

The fish curry was delicious. I told Fabrikage that and he was euphoric.

'What is this fish called?' I asked. 'I don't recall having ever eaten this fish before.'

'It's called silver belle,' Fabrikage said. 'It's very popular in these islands.'

'What are the other popular varieties in the local market?' I asked.

'Well, sardine is popular,' Fabrikage explained. 'So is mackerel.'

'What about tuna and shark?' I asked.

'We get them too,' he said. 'But in the tuna category, what is popular is the little teenie.'

We finished our very satisfying lunch and went down the steps to admire the sea lashing at the compound wall of the bungalow. We sat down on the cement bench thoughtfully placed in that vantage position, as if strictly for visitors like us who can never have enough of the sea. After a while, I started talking to Dr Shailesh Nayak about how satellite observation helps in increasing the fish catch.

'The basic idea,' Dr Nayak explained, 'is to understand how fish migrate in marine waters for feeding. The task is one of identifying the feeding grounds.'

'How do you do that from a satellite?'

'Well, upswelling brings the nutrient-rich cooler water to the surface of the ocean. That enhances biological activity. This shows in the form of anomalies/gradients in the sea surface temperature pattern.'

'So what happens?'

'A satellite has thermal sensors. These sensors can detect these anomalies/gradients in the sea surface temperature pattern. So the data from these thermal sensors are used as a clue to demarcate areas where fish aggregate. But sometimes, there can be a problem with this approach.'

'What would that be?'

'You see, the sea surface temperature data generated by the satellite provide only surface information on water circulation. Suppose, during summer, the sea surface gets heated up, as it happens in the marine waters.'

'As in the Indian marine waters, isn't it?'

'Exactly. Then, it gives rise to strong stratification of the water mass. That prevents the arrival of the cool nutrient-rich water from the deeper layers to the surface. In that case, the satellite sensor does not detect the sea surface temperature gradients.'

'What do you do to address the problem?'

'There is another way. You see, phytoplanktons are microscopic plants, which occupy the first link in the marine food chain. Phytoplankton contains photosynthetically active pigment which is called chlorophyll-a. The ocean colour monitor of the satellite can detect the concentration of chlorophyll.'

'So what do you do?'

'We integrate the chlorophyll concentration and the sea surface temperature images and make the forecast for PFZs.'

'What is a PFZ?'

'PFZ stands for Potential Fishing Zone. In other words, the prospective sites for fish accumulation. The technique for PFZ forecast using integration of sea surface temperature and chlorophyll-a has been transferred by ISRO to the Indian National Centre for Ocean Information Services, INCOIS for short.'

※※※

Dr Koya arrived to take us to his office. In his office, we talked about the PFZ advisories.

'We get the PFZ advisories thrice a week,' Dr Koya said. 'On Tuesdays, Thursdays and Saturdays.'

'How do you get them?'

'We get them by fax from INCOIS. Nowadays, INCOIS has a website which we use in case there is some difficulty with the fax.'

'Do you pay for the PFZ information?'

He shook his head. 'No, it's all free. We, in turn, supply the information free to the local fishermen.'

'Do you have any problems accessing the PFZ advisory?'

'Sometimes. This happens particularly when the sky is cloudy.'

'What's the language of the advisory?'

'Previously we used to get it in English. But now, it is in Malayalam. But since all the fishermen here are educated, we don't have any problem whether it is in English or Malayalam.'

'Tell me something, Dr Koya,' I asked. 'Do you think the PFZ advisories have been useful to the fishermen here?'

He nodded. 'Yes, of course. They've been very useful. As a result of these PFZ advisories, the fish landings have improved substantially. The only problem, as I can see, is that the PFZ information is not very regular.'

Dr Shailesh Nayak intervened to say, 'But do you know why it is so? That's because PFZ information can be provided only when the satellite provides cloud-free data. If there are clouds covering an area, there's nothing much ISRO can do.'

'I do understand,' Dr Koya said with a broad grin. 'But let me assure you that these PFZ advisories have been extremely useful to the fishermen here in the islands. In any case, we're now going to meet some of them. You can talk to them yourself and find out.'

※※※

We met them in the hall across Dr Koya's office. There were several of them; Dr Koya had been diligent, I thought. They were Andar, Hamsath, Jaffer, Kashmi, Rafeeque, Ali, Khaleel, Ali Akbar, Muthu, Haneefa Koya, and Abdul Khader.

After the introductions were over, Abdul Khader spoke up, 'Dr Ismail Koya has told us why you are here. We're happy that you people have come so far to find out how the scheme has worked.'

'How has it worked?' I asked him.

'Very well,' he said with an emphatic nod of his head. 'The information, I must say, is very useful. Dr Ismail Koya gets it from . . . I know it is from Hyderabad, but I've forgotten the name of the office that sends the information.'

'INCOIS,' Dr Ismail Koya said.

'Yes, that's what it is,' Abdul Khader said. 'As soon as Dr Koya gets the information, he passes it on to us. All I can say is that the information has increased our fish catch.'

'In what way?' I asked.

It was Jaffer who chose to answer my question. 'Well, let me tell you what used to happen before. We used to go long distances in the sea in search of shoals of tuna. Sometimes, we came back without fish. Such a waste of time it used to be!'

Andar chipped in, 'And not to speak of the burden of diesel and other expenses on such fruitless trips! Believe me when I say that the burden was quite heavy.'

Nodding in affirmation, Hamsath added, 'Yes, that's true. It was a heavy cost. Both in terms of time and money. Now that Dr Ismail Koya gives us the information, we know exactly where the shoals of tuna are going to be. What do you call this information, Dr Ismail?'

Dr Koya smiled indulgently at Hamsath. 'PFZ advisory. Can't you remember this simple thing?'

Hamsath grinned sheepishly, 'I'll remember it henceforth. How my mind wanders!'

I asked, 'Do you people know how to read these PFZ maps?'

It was Rafeeque who answered. 'Yes, we do. You see, it's quite easy really. In addition, Dr Ismail Koya has taught us how to read these maps correctly so that we don't make mistakes.'

'But you haven't told us how you read these maps?' Dr Shailesh Nayak asked Rafeeque in the vexed tone that teachers normally use when dealing with recalcitrant students.

Rafeeque now wore an expression of injured innocence. 'I thought I told you. It's quite easy. The PFZ map gives a scale so that we can measure the distance in kilometres. And the map shows the area where the shoals of fish are going be. It is in thick black colour. It's fairly easy to identify these areas.'

Ali Akbar who was quiet all this time, intervened to say, 'And now the map is written in Malayalam. So that anyone here can read the map.'

'That's because you're such a dunce,' Rafeeque said. 'As if fishermen here had any difficulty reading the map when it was in English!'

Kashmi spoke for the first time. 'You are being unnecessarily harsh on Ali Akbar, Rafeeque Mian. Isn't it a good thing that the map is written in Malayalam? I know there are many people in Androth Island who can't read English.'

Jaffer intervened, 'Why should we quarrel about what language the map is written in? There are many fishermen in the mainland who can't read a map even if it's written in the local vernacular! But there is one thing good about these maps.'

'What is that?' Dr Nayak asked.

Jaffer took his time in answering Dr Shailesh Nayak's question. He said after a while, 'The good thing is that the map shows on which day it has been issued. And till what day the forecast is valid.'

Abdul Khader nodded sagely. 'What Jaffer says is right. Since the map gives dates, we don't roam around the sea with old maps which are no longer valid.'

'Has the tuna catch gone up because of these satellite maps?' I asked.

Abdul Khader said, 'Yes, yes. Most certainly! Our catch has gone up.'

Khaleel spoke up now, 'Tuna catching is not the same as catching any other fish. We have to cook and smoke the fresh tuna immediately after catching it. And then dry it in the sun. And then only it becomes *Mas* and can be sold. All this takes time!'

'I know it takes time,' I told Khaleel.

Khaleel smiled at me. 'Well, what I was trying to say is that because making *Mas* takes time, we need to save time while searching for the shoals of tuna. And that's where your information helps.'

Muthu had been quiet all this time and biting his nails. He intervened now to say in an animated voice, 'Well, Khaleel is right. We save a lot of time now because the maps tell us where the shoals of tuna are. And let me tell you that the maps are always right. We find the shoals of tuna where the maps say they are going to be.'

'So you save time because of these maps,' I pointed out. 'But what do you do with the time you save?'

Muthu gave me a smile, a sheepish smile. He said, 'We now spend that time with our family. Do you know what used to happen before?'

'Tell me,' I said.

'We hardly had any time to be with our family. We used to set off fishing before daybreak. Sometimes, even earlier! By the time we came back home after cooking, smoking and drying whatever tuna we had caught, it was late at night and people at home were fast asleep!'

'Not to speak of what a grind it was!' Khaleel said with a frown. 'Day after day!'

Haneefa Koya spoke now, 'That's true. What a horrible life it was! But even in spite of all that work, we hardly got any tuna. Our search for tuna in the big ocean was like that story I heard the other day. About searching for a needle in a haystack.'

He chuckled loudly at his own joke. The chuckle became

a paroxysm of laughter. With each guffaw, his chins rippled, and his belly rose and fell. Finally he stopped.

'Did you see how Haneefa Koya laughed?' Khaleel asked. 'That too, at his own joke! Well, that's why he never catches any fish. The shoals of tuna run away when he guffaws like that!'

It was Muthu who spoke now. 'But what Haneefa Koya said was correct. In those days, the fish catch we got was very little. Hasn't the catch improved after the maps are available?'

'Yes, that's true,' Abdul Khader told me. 'We've benefited from your maps. But the problem is that the supply of your maps is not regular. We suffer now because we can't get the maps regularly.'

I decided that Dr Shailesh Nayak should field Abdul Khader's complaint about the supply of maps being irregular. I requested him to respond to the question.

'What Abdul Khader says about the maps not being available sometimes is true,' Dr Nayak explained. 'But that happens when there are clouds in the sky. And ISRO's present satellites can't look at the ocean when the clouds are there. That's why we can't supply maps to you at that time.'

'But when is ISRO going to make a satellite that can see through the clouds?' Dr Ismail Koya asked.

'Very soon,' Dr Shailesh Nayak assured them.

※※※

Our meeting with the fishermen was over. I thanked them for talking to us so freely and we drove back to the bungalow. Dr Ismail Koya suggested we could have our tea at our favourite place overlooking the sea. There were two large vessels honking away in the channel.

'What are these vessels?' I asked.

'Oh, Khadeejah Beevi and Hameedath Beevi,' Dr Koya said. 'They are the inter-island ferry vessels. Catamaran-type, high speed vessels. They connect the islands.'

The waves were bigger now, almost menacing. They crashed into the shore near the bungalow with a wild hiss, spitting froth in an ever-widening swath. Dr Koya said high tide was starting.

Dr Koya gazed intently at the horizon and declared that the fishing boats were coming back. I looked at the horizon, but all I could see was watery starkness: water heaving, seething and foaming like churned milk.

'I can't see any fishing boats,' I said.

Dr Koya asked me to shade my eyes with my palms and look again. I could now make out small specks on the rim of the horizon, wriggling like tiny water snakes. Fabrikage brought our tea and served it to us.

The wind turned wicked now, blowing sand into our eyes. It howled, sounding like a million flutes singing out of tune. It whipped around the trees in the compound of the Inspection Bungalow and the tops of the trees swayed wildly, as if in a frenzy.

The fishing boats came now, gliding through the waves. All of a sudden, white-breasted gulls descended; they came swarming and screeching, and whizzed past so close that I could see their eyes: glassy and immobile like the eyes of a lizard. They swooped down with a swish and I raised both my hands to protect my face.

Dr Koya laughed uproariously, as if hugely amused.

'Don't worry,' he said, putting a comforting hand on my shoulder. 'They'll do nothing to us. They are only after the fish that the fishing boats are bringing in.'

'Aren't you happy that the fishing boats are coming back so early?' Dr Nayak told Dr Koya. 'That way, the fishermen will have more time to spend with their family.'

'Yes,' Dr Koya said. 'All thanks to your PFZ advisories. But you must do something about the regularity of sending the PFZ information.'

I looked meaningfully at Dr Shailesh Nayak, but made a mental note of talking to Dr Radhakrishnan, the head of INCOIS about it. I called him that night, but he was away at Agra attending a conference.

※※※

By the time I could manage to meet Dr Radhakrishnan at Hyderabad, a few weeks had elapsed. I met him in his office at INCOIS. Dr Radhakrishnan, who was one of the leading application scientists of ISRO, was currently with INCOIS. Lean of built, he has a luxurious head of hair, piled up high in the style of a movie star and a beard to match. He is very articulate, but when he speaks, different parts of his body move in a rhythm as if to emphasize the point he is making. Not surprising, considering that he is an accomplished Kathakali dancer.

'I've checked the position about the supply of PFZ advisories to Lakshadweep Islands,' Dr Radhakrishnan said. 'It has been irregular at times, but only during the rainy season. That's because there are too many clouds in the sky over Lakshadweep Islands at that time of the year.'

'I know,' I said. 'But the islanders made a point of complaining about how irregular the supply of PFZ advisories has been. But they also said how greatly they had benefited from the advisories.'

His face brightened. 'Did they?' he asked. 'Well, that seems to be the experience of all the users to whom we provide the PFZ advisories.'

'Tell me,' I asked. 'What all are you doing at INCOIS to help the poor fishermen?'

'Everything in our power,' he said with that precise, decisive gesture that made him such an accomplished Kathakali dancer. 'We've now taken up this work of issuing PFZ advisories on a mission mode. At present, we disseminate the advisories to active nodes spread around the entire coastline of India.'

'How many active nodes do you have now?'

'We've recently increased the number of nodes where we disseminate information from 110–200.'

'As many as that?' I said. 'Tell me, where are these nodes located?'

'We've now provided nodes at all the offices of the state fisheries departments, at the regional offices of the Central government agencies, at public sector undertakings and at the fishermen associations. We also provide the PFZ advisories to individual fishermen who specifically request for information.'

'Well,' I observed, 'the spread of the nodes is really impressive. Have you also made other improvements?'

He nodded. 'Yes, we've made several improvements. For example, we've improved the timeline for processing and disseminating the information. At the same time, we have made sure that the dissemination of information is done at fixed slots.'

'That'll certainly help,' I said.

'You see,' Dr Radhakrishnan continued, 'we used to issue these advisories in English. But now we issue them in the local vernacular.'

'That has helped,' I said. 'I saw that for myself in the Lakshadweep Islands.'

'We've also taken up extension programmes, so as to make the users aware of the usefulness of the advisories that we provide. We've conducted user awareness workshops at Ratnagiri, Nizampatanam, Kakinada and Mumbai. We're planning to conduct some more in the coastal states.'

'Are these workshops useful?'

Dr Radhakrishnan nodded. 'Yes, they are useful. They create the necessary awareness amongst the users. It's also necessary to educate the functionaries of the fisheries departments of the state governments because they provide that crucial link with the users in the coastal areas.'

'Have you done anything towards that end?'

'Yes, yes. We've conducted training programmes for the officers of the fisheries departments of Andhra Pradesh, Tamil Nadu and Pondicherry. From the feedback that we've got at INCOIS on these training programmes, they have been hugely successful.'

'That's very commendable,' I said. 'Tell me, what about the dissemination of the PFZ advisories? How do you do it?'

'Well, we do it by fax, telephone, digital display boards and posting the information on the INCOIS website. We've put up information kiosks and digital display boards at all fishing harbours. We've also increased the frequency of the forecasts. Previously it used to be twice a week. We've now increased it to thrice a week, on Tuesdays, Thursdays and Saturdays.'

'That's true,' I said. 'I was told about it in Kavaratti. Let me ask you a general question. Do you think, with all that you're doing, there has been an increase in the fish catch?'

'Oh, yes. You see, we've done some observations in Gujarat state for the fishing strata classification of depth zones of 30–50 metres, and 50–100 metres. We have something we call normalized Catch per Unit Effort (CPUE). What we did was to compare the CPUE of each observation with the normalized CPUE of the fishing vessel for each month.'

'What did you find?'

'Some observations indicated abrupt increase in catch, two to threefold in the PFZs. The per cent increase in total catch was calculated from CPUE in PFZs as compared with the mean CPUE of the month. We saw from the feedback that there was about 100 per cent increase in catch in PFZs located in 30–50 metre depth zone and 70 per cent increase in catch in the PFZs located in the 50–100 metre depth zone.'

'That's very impressive,' I observed. 'Are there other feedbacks?'

'Yes. According to the concurrent validation of PFZ advisories done for the Kerala coast, it was found that the searching time for fish has reduced by 30 per cent to 70 per cent.'

'That's very good,' I said. 'But do you think, the life of the ordinary fisherman has improved from all this?'

'Most certainly,' Dr Radhakrishnan said. 'Let me tell you something. The PFZ mission is perhaps one of the best examples of translating the fruits of space technology to benefit the common man.'

I smiled, realizing that though it sounded strange, space technology was having a deep impact on an industry as far removed from it as fishing.

5

IN TRIPURA
Providing Healthcare

Udaipur is a small town in the state of Tripura, with a maze of brightly coloured temples and winding lanes. Throughout the day, this pilgrimage town throbs with activity: pilgrims chant loudly, cymbals clash, conch shells sing, trumpets blow and bells clang. Loudspeakers, hitched to treetops, quote from the scriptures. Cramped shops shriek at each other as if to outdo the pilgrim babble. Standing tall in that deafening noise is the most famous temple of Tripura, dedicated to Tripura Sundari, the presiding deity of the region. The goddess' name graces all buildings of note, even the government hospital.

Satyamurthy and I were in Udaipur to look at how the telemedicine connectivity in the Tripura Sundari District Hospital was working. Satyamurthy, an engineer who built satellites all his life, was now the head of ISRO's telemedicine programme. He has a sense of humour and a deep voice. Small wonder, then, that he is the natural choice to speak at all ISRO functions where he entertains the audience with his stock of jokes and funny one-liners delivered in his rich baritone. Before we left for the hospital, I made Satyamurthy tell me about ISRO's telemedicine programme.

'The idea of telemedicine,' Satyamurthy explained, 'is to provide healthcare from a distance. The telemedicine system has two ends. One is the specialist end, and the other, the patient's end. The specialist doctor advises the patients, the non-specialist doctor, or even the paramedic at the patient's end, while online, about medical care.'

'That's great,' I said. 'Tell me, what does the telemedicine system consist of?'

'Three things, essentially. One is the customized medical software, second is the computer hardware and third, the medical diagnostic equipment. The medical software is integrated with the computer hardware along with the diagnostic instruments connected through satellite-based communication.'

'How are they connected to the satellite?'

'Through VSATs with antenna and transmitters at each location. We control it from an ISRO hub station that is centrally located. The hub station provides the network monitoring and control, serviced through bandwidth from ISRO satellites.'

'The idea is to provide the facility for transmission, isn't it?'

He nodded. 'Yes, the facility for transmission of the patient's medical images, records, outputs from medical devices in addition to providing live two-way audio and videoconferencing. The medical record or history of the patient is sent to the specialist doctors in a faraway place, either in advance or on a real time basis. The specialist doctor studies the medical history of the patient and provides the diagnosis and treatment during the videoconferencing.'

'So essentially, ISRO provides the connectivity between the patient's end and the specialist's end.'

'Yes, that's true,' he said. 'You see, when we started on the telemedicine project, the connectivity was limited from one point to another point. What this meant was that one patient end was connected to one specialist doctor at the super speciality hospital.'

'Why point to point?' I asked. 'Was there a bandwidth problem?'

'No,' he said, his voice gaining a higher pitch. 'That was because we were experimenting. Soon we got around to making improvements. That meant that we could connect one patient at a time to any of the specialist doctor's end within the super speciality hospitals.'

'Any further improvement?'

Satyamurthy nodded. 'Yes. We now have a multi-point system. That means that several patient's ends are simultaneously connected to different doctor's ends at different hospitals at different geographical locations.'

'That gives a much greater reach, doesn't it?'

'Yes, it does.' Satyamurthy said, his voice exultant. 'The important thing about telemedicine is that it is a programme for the common man that utilizes the gains made in different fields of science and technology.'

'Why do you say that?' I asked.

'You see,' Satyamurthy said slowly, like a teacher explaining a difficult sum, 'telemedicine is the confluence of communication, information technology, bio-medical engineering and medical science.'

'Well, nowadays, I am told telemedicine can be done through telephone or emails. How is ISRO different?'

'There is a difference. You see, space-based telemedicine is interactive.'

It was time for us to make our way to the hospital.

✕✕✕

We decided to walk down to the hospital. It was located in the heart of the town, and I noticed that like most government hospitals, it was not particularly clean. The walls were dirty and the plaster was peeling in the corridors and rooms.

We were expected. The chief administrative officer took us around the hospital and finally deposited us in the room of Dr Kanu Lal Bhowmik, the medical superintendent of the hospital. Dr Sibas Das, the doctor in charge of the Coronary Care Unit (CCU) and Dr Tapas Dutta, the medical officer of the CCU were also there.

Dr Bhowmik explained to us the activities of the hospital. 'We have ten doctors, forty-one nurses, five pharmacists, eight technicians, and forty-two medical assistants. The hospital has facilities for both inpatients and outpatients.'

'How many outpatients visit the hospital every day?' I asked.

'On an average, about 350 outpatients,' Dr Bhowmik said, in a gruff voice.

'What about inpatients?' Satyamurthy asked.

'We have facilities for them, too,' he explained. 'We have 135 beds.'

'But what I saw, Dr Bhowmik,' I pointed out, 'is that the inpatient ward was very crowded. I could count at least fifteen patients sleeping on the floor.'

'That happens,' he admitted sheepishly. 'The beds are limited. So many patients need to be admitted; we can't turn them away, can we?'

Dr Sibas Das intervened to say, 'This happens in all the small town hospitals.'

'That's true,' Dr Bhowmik said. 'Well, I believe you people are interested in watching our telemedicine programme. Dr Dutta will take you to the CCU and show you around.'

We accompanied Dr Dutta to the Jogesh Chakroborty Coronary Care Unit, which is the telemedicine centre. The floors were covered with good quality tiles and the walls were painted a shade of soothing blue, which matched the colour of the curtains. The whole of the CCU was air-conditioned.

'The CCU has six beds,' Dr Dutta explained. 'There are three doctors here. They've been given preliminary training at the Rabindranath Tagore International Institute of Cardiac Sciences (RTIICS), Kolkata. The idea is to stabilize the patient and give him preliminary care.'

'What about paramedical staff?' Satyamurthy asked.

'We have four Cardio Vascular Technicians (CVTs) who

deal with the technical part right from taking the ECG to transferring the data,' Dr Dutta said. 'There are nurses and attendants also.'

'Which super speciality hospital are you connected with?' I asked.

'You see, our hospital is the patient's end,' Dr Dutta explained. 'We are connected to two super speciality ends. One is the RTIICS, Kolkata, and the Narayana Hrudayalaya, Bangalore.

'What exactly do you do?' I asked.

'In case of inpatients,' he said, 'as soon as a patient is admitted to the CCU, his medical data is instantly sent to RTIICS. The specialist doctor at the other end provides instant diagnosis. Let me tell you how it's done. When patients with chest pain report at the CCU, the medical officer examines and takes down the history of the patient. ECG is taken within five minutes by the nurse in the CCU, assisted by the CVTs. ECG along with patient history is then transmitted to the specialists in RTIICS within ten minutes by CVTs.'

'That's fast work,' I said. 'What does the specialist at RTIIC do?'

'The cardiac specialist there confirms diagnosis and advises treatment,' Dr Dutta explained. 'The medical officer here follows-up with the advice. We conduct videoconferencing if there are complications or if the specialist wants to check-up facts with the patient. That's done quite often.'

'That's for the inpatients,' Satyamurthy observed. 'What about the outpatients?'

'The cardiac patients from the Out Patient Department (OPD) report to the CCU,' Dr Dutta said, his voice rising a few decibels. 'The doctors and the CVTs check old reports and take down the medical history. You see, the telemedicine consultations in the videoconferencing mode for the outpatients take place on Mondays, Wednesdays and Fridays from eleven to twelve noon.'

'What happens in the telemedicine consultations?' Satyamurthy asked.

'Advice of the RTIICS specialists is handed down in the prescribed proforma to the patients. And we do the follow-up. But emergency patients get emergency attention. Both here and at the RTIICS.'

'When do the telemedicine consultations start today?' I asked.

He looked at his watch.

'Any time now. Today is earmarked for outpatients. And we're lucky today because Dr Devi Shetty from Narayana Hrudayalaya, Bangalore, is visiting RTIICS today. He has agreed that he would be available for teleconsultation for the outpatients here today.'

He looked at his watch again.

'I don't know what's holding up the consultations. Let me go and telephone RTIICS.'

The patients for telemedicine consultations had already arrived and taken up positions in the room. They looked poor. We caught hold of a CVT who was shepherding the patients around and engaged him in conversation.

'Where are these patients from?' I asked.

'They are from all over Tripura,' he explained. 'Some of them have even been referred from GB Pant Hospital, Agartala. There are some from Bangladesh also.'

'They look poor, don't they?' I suggested.

He made a clucking sound. 'Most of the patients who come here for telemedicine consultations are poor. They are certainly below the poverty line. In any case, they are mostly illiterate and some of them are semi-literate.'

'Do you charge them for the consultation?' I asked.

He shook his head. 'No, we don't. Patients below the poverty line are treated free of cost.'

'Is telemedicine consultation useful to these patients?' Satyamurthy asked.

He nodded. 'Yes, very useful. For these poor people who otherwise can't consult a cardiac specialist without travelling miles, this facility is life-saving. It gives them a chance at living, surviving.'

'Why do you say that?' I asked.

He said, waving a hand dismissively, 'These patients here are simple people who don't even get a first opinion, leave alone a second opinion. Do you know something? Many patients who come here don't even know that they are the beneficiaries of telemedicine facility.'

'Is that so?' I asked.

He allowed himself a thin smile. 'Absolutely. The patients here in Tripura are village people who regard doctors as demigods. Sometimes, of course, a politician comes here, who talks big and drops names. But the general attitude of the patients is awe and respect.'

'Why should it be so?' Satyamurthy asked.

He wagged a finger at Satyamurthy as if scolding a child. 'That's because the educational level of most of the patients is not very high. They would have done Senior School Leaving Certificate (SSLC) or High School Certificate (HSC). But let me tell you one thing. Their memory, as far as their ailment is concerned, is absolutely clear. Most of them talk about their problems, dates of visit to doctors and treatment in perfect chronological order.'

'That's rather good for people who are not highly educated,' I observed.

He nodded. 'Isn't it? For patients living in remote areas like Tripura, talking to specialist doctors in Kolkata or Bangalore through television is a miracle that they still can't believe.'

'Why is that?' I asked.

His voice rose. 'Because it gives hope to all these patients who had given up hope of living a healthy life due to lack of resources and appropriate medical facilities.'

'How are you so familiar with all these?' Satyamurthy asked.

Pride coursed in his voice. 'I should know, shouldn't I? I have been working as a CVT in this telemedicine unit ever since it was inaugurated.'

'When was that?'

He cleared his throat. '24 June 2001 was the date. I remember the date very well because I was there. Since then, thousands of patients have had the benefit of being treated by super specialists, thanks to the telemedicine connectivity.'

'What do you think has been the major benefit for the patients from the telemedicine connectivity?' I asked him.

'The major benefit for the patients in Udaipur,' he declared, stroking his chin thoughtfully, 'is that they get specialized treatment in times of need as a result of which they can recover quickly. Due to prompt communication, there is better treatment for the patients.'

※※※

Dr Sibas Das walked in and the videoconferencing started almost immediately. The TV at the end of the room spluttered for a few seconds before coming to life. The images on the TV were from RTIICS, the super speciality end at Kolkata. A telegenic face appeared on the TV and the patients in our room burst out clapping. The clapping lasted a few minutes.

'That is Dr Devi Shetty,' Dr Dutta, who was sitting next to me, explained with pride in his voice, 'the world renowned heart surgeon.'

'But why are the patients clapping so enthusiastically?' I asked.

'He is like a god to everyone here,' Dr Dutta replied exultantly. 'Very popular. He is the poor man's heart specialist. These people here think he can cure any heart ailment!'

The consultations started. It was the turn of the first patient. Dr Devi Shetty looked at the medical records of the patient which had been sent to him earlier and asked Dr Das about the general condition of the patient: whether he had any extra complication and the type of medicine he was taking. He now talked to the patient in broken Bengali and told him that he was all right. All he needed to do, Dr Shetty told him, was to take his medicine regularly and avoid oily food.

The next patient, according to Dr Shetty, required surgery. He informed the patient about the approximate time and cost of the surgery. He asked the patient whether it would be possible for him to come to Bangalore to have the surgery.

The next one was a two-year-old boy who, Dr Shetty felt, needed surgery. He talked to the father of the patient about his economic condition and declared that he would do the surgery free of cost.

Next was the turn of an eleven-month-old baby, who slept on its mother's lap, blissfully unaware of the consultations. Dr Shetty studied the medical history of the patient, talked to Dr Sibas Das and the baby's mother and declared that the baby did not need any surgery rightaway. Costly surgery, he said, need not be done only for the heck of it.

Dr Shetty talked to all the patients who had come in for telemedicine consultations one by one. By the time he had finished, the telemedicine consultations had lasted the best part of two hours. The consultations ended and there was a round of loud clapping from the patients.

There was something very endearing about how Dr Shetty conducted the consultations. He spoke to the patients in broken Bengali which increased the patients' sense of oneness with him. He would stop in the middle and ask the consulting doctor a few questions regarding the medical condition of the patient, his medication and provide diagnosis on whether the patient required surgery. Dr Shetty would then ask the patient whether there was anything else he would like to know.

This personal interaction erased, for most part, the impersonality of the teleconsultations. Kudos to Dr Shetty for managing to create a personal rapport with the patient, sitting 1000 kilometres away, I thought.

※※※

I started talking to some of the patients to find out how they felt about the telemedicine consultations. I talked to the patient who had been advised by Dr Devi Shetty to eschew oily food.

'How do you feel about the telemedicine consultation we just had?' I asked. 'Was it any good?'

He nodded enthusiastically. 'Of course, it was good. And very productive, too.'

'Why do you say that?'

'Well, the people here,' he gestured expansively at the patients who had participated in the telemedicine consultations, 'are poor people. Can any one of us consult a specialist without travelling all the way to a very big city? Or, for that matter, without spending a lot of money? In any case, the question doesn't arise because we don't have that kind of money! That's why I always tell people that a thing like the telemedicine is life-saving for folks like us who're poor and live a long distance away from big cities.'

That obviously was a long speech by his standards and I could see that the effort had tired him. I waited for a while before speaking again.

'But all the advice that Dr Shetty gave you was not to eat oily food!'

He said, a grin spreading across his face, 'That's true. But if Devi Shetty says I don't need surgery, then I don't need surgery. There isn't a more qualified person in the entire world to advise me! Let me tell you what happened to one of my relatives. Right here in this hospital. You see, this relative of mine is an old man. He complained of very acute pain in the chest. So we brought him here to this hospital in the dead of the night.'

I grimaced. 'What a time to bring a sick man to a government hospital! I hope somebody was there to at least open the door.'

He said, with a bemused gesture, 'We were lucky! There was a doctor around at that time of the night. The doctor checked my relative and said it was a case of total heart blockage. What a thing to happen!'

'So what did you do?' I asked.

'All hell broke loose,' he said, clucking his tongue in annoyance. 'Every few minutes, his heartbeat would become highly erratic, and the doctor and all those young boys with him would rush in. By the way, what do you call those young boys with the doctor?'

'Cardio Vascular Technicians,' Satyamurthy explained.

'Whatever!' he said, shrugging his shoulders. 'Those boys would kneel down and give my relative a cardiac massage. That would revive him. But only temporarily, mind you!'

'What happened, then?' I asked.

His voice had an edge to it. 'We begged the doctor to do something more permanent to save my relative. But what could the poor doctor do? You see, my relative needed a pacemaker. Otherwise, the doctor told us, there was no chance for him to survive the night.'

'Why didn't they put a pacemaker?' I asked.

He looked exasperated. 'You are forgetting, sir, that this is only a small, moffusil hospital! They don't have that kind of facility here! That could be done only at Kolkata. And my relative was certainly not in a position to travel all the way to Kolkata.'

'So that was the end of your relative, was it?' I asked.

He shook his head vigorously. 'No, no. Nothing like that! The doctor here did something very sensible. He sent all the medical records of my relative to the RTIICS in Kolkata on the telemedicine line, and consulted a specialist there. Smart thing to do, isn't it?'

I nodded in agreement.

'The specialist in Kolkata,' he continued, 'prescribed some medicine to be given immediately to my relative. The medicine was given. And would you believe it, the condition of my relative improved dramatically?'

'Did your relative survive finally?' I asked, my voice a trifle impatient.

'Oh, yes, he did,' he said, chuckling with satisfaction. 'Next morning, my relative was up in his hospital bed and talking to us like any other person does. Like a normal person, I mean. I can tell you that it was nothing short of a miracle. Don't you agree?'

'Yes,' I said.

'It was, it was!' he insisted in a vigorous tone. 'You see, after that incident with my relative, I tell people that telemedicine saves lives.'

※※※

Next, we went to talk to the mother of the eleven-month-old baby who had been examined by Dr Devi Shetty on the wire.

'What's his name?' I asked the mother, gesturing at the child.

'Subhodip,' she said bashfully.

'That's a lovely name,' I said. 'Aren't you happy that Dr Shetty said the baby didn't need surgery rightaway?'

She nodded eagerly. 'Yes, yes. You know something? I have been coming regularly for telemedicine consultation with Subhodip. Today is the third time.'

'Why, what happened?' I asked her.

She clucked her tongue. 'You see, Subhodip was in great pain. He was always crying away in pain. He wouldn't go to sleep. So I brought him here.'

'What did the doctor here say?' Satyamurthy asked.

She pointed to Dr Tapas Dutta and said, 'That doctor there examined Subhodip. He said, "This baby has a heart problem." Do you know what they did? They sent Subhodip's medical records to that big hospital in Kolkata.'

'Rabindranath Tagore International Institute of Cardiac Sciences?' I asked.

She brightened. 'Yes, yes. That's the name of the hospital. That was the time when we had the first consultation on the telemedicine line.'

'What did the specialist doctor say?' I asked.

She wrinkled her brow. 'Well, the specialist asked me several questions about Subhodip's condition. He looked at all the medical records.' She pointed at the telemedicine TV. 'Do you know something? I could see it all on that TV!'

'How did it feel?' Satyamurthy asked.

She smiled at Satyamurthy. 'I felt as if the specialist was right here in this room talking to me! He asked me to give some medicine to Subhodip. He also said I must bring Subhodip again after a month for telemedicine consultation.'

'Did you do that?' I asked.

She nodded. 'Yes. But Subhodip was already feeling much better after the medicine. But when the specialist looked at Subhodip's condition after a month, he felt that a surgery might be necessary.'

'Immediate surgery, is that what the specialist said?' I asked.

She shook her head. 'No, no. Not immediate. He said some more telemedicine consultation was necessary. He himself had given today's date for consultation.'

'Are you happy with today's telemedicine consultation?'

She smiled. 'Yes, very happy. That specialist from Bangalore was very good. Wasn't he?'

I nodded. 'He is Dr Devi Shetty.'

Her face lighted up. 'He is kind, too. Didn't you see that? He has now given me so much hope for Subhodip. This thing,' she pointed to the telemedicine TV, 'is a real miracle. It is as if these big doctors are here in this room and giving you advice. Yes, certainly, for poor people like us, it is a real miracle!'

'Yes, yes, it's a miracle,' said a voice from the back. I turned around to see. It was the person in whose case Dr Shetty had suggested surgery.

'Why do you say that?' I asked him.

'You see, I am from Bangladesh,' he explained. His cheeks were very plump, like the swollen cheeks of a bugler in the middle of a tune. 'I came here when I heard about this telemedicine facility in Udaipur. I needed the services of a specialist.'

'Why, what's your problem?' I asked.

He put a hand on his chest. 'I've a very bad heart. And it was my great good luck that it was Dr Devi Shetty on the television today.'

'How did you hear of Dr Shetty?' I asked.

He said, a smile wrinkling his face, 'Oh, he is a legend! He has operated on so many patients in Kolkata and they all got well. Can you really believe that I talked to him today? And it is all because of that miracle,' he said gesturing towards the TV.

'Is that so?' I asked.

He nodded. 'Of course, yes. Dr Shetty will operate on me and I'll get well. I am a poor man and I'll tell him that. He'll give me a concession. Maybe, he'll do it free. They say he is very kind to poor people.'

※※※

After talking to the patients, we went back with Dr Das and Dr Dutta to Dr Bhowmik's room for the promised cup of tea. Dr Bhowmik was happy to see us.

'So how did the consultations go?' he asked in his usual gruff voice. 'I believe Dr Devi Shetty was in the consultations today. He's such a big hit with the patients, isn't he?'

I beamed a smile at Dr Bhowmik. 'He certainly is. The patients treat him as if he is some kind of a divinity.'

'I suppose so,' Dr Bhowmik said. 'But he also happens to be one of the most outstanding heart surgeons of the country today.'

'Let me ask you a question, Dr Bhowmik,' I said. 'What do you think of telemedicine?'

'I think it saves lives and time,' Dr Bhowmik said. 'But let Dr Das and Dr Dutta tell you that. After all, they are the doctors who've hands-on experience of the telemedicine programme. What do you think of it, Dr Das?'

'I think it is an excellent programme,' Dr Das said. 'My views are very positive. Telemedicine has provided increased accessibility, specialized medical consultation, second opinion, expert opinion as and when required and during emergency situations in remote places.'

'My views are positive, too,' Dr Dutta said with a smile.

'I think in a remote and inaccessible place like Tripura, it is of great help. Because of timely treatment in case of emergencies, there is minimization of severe damages, which would have happened if the patient had to be taken to a faraway place like Kolkata.'

'Tell me, something,' I asked. 'Do you doctors who are at the patient's end, feel threatened by the invasion of specialist doctors that is made possible by the telemedicine system?'

Dr Das shook his head. 'No, we don't. We recognize the fact that telemedicine has helped patients in interior and far-flung places to get the benefit of specialist opinion.'

'That's not what I meant,' I said. 'You see, telemedicine introduces technology into the practice of doctors. The doctors may feel threatened by that. In addition, they may feel that the personal relationship between doctors and patients might suffer. Do you agree?'

It was Dr Dutta who replied. 'What you are talking about is the resistance to change when new technologies are introduced. But let me tell you that telemedicine is so useful that I don't see any resistance taking place in the case of healthcare professionals.'

Dr Bhowmik intervened to say, 'No, we don't feel threatened. On the other hand, telemedicine provides opportunities for us to consult other doctors almost instantly for clinical diagnosis. That helps us. There's something else.'

'Like what?' I asked.

He said, stabbing the air with a finger, 'You see, this facility of keeping in touch with other fellow professionals helps us to enhance our academic knowledge. It also enables us to keep track of new breakthroughs in our particular fields.'

'That's true,' I said.

'There's another way in which telemedicine has helped,' Dr Dutta said. 'Because of telemedicine, we now keep a database of patients and their ailments. This is available immediately in case of future references and emergencies.'

We soon took our leave because we had to travel to Agartala and take a flight from there to Kolkata.

※※※

On the flight to Kolkata, I started talking to Satyamurthy.

'Well, your telemedicine connectivity seems to be such a great boon to the poor people,' I pointed out. 'Looks like it is the only way for these helpless fellows to consult a specialist. That is at least true of Udaipur. What's your experience of other places where ISRO has provided telemedicine connectivity?'

'Equally good,' Sathyamurthy said. 'Let me tell you what happened in the Lakshadweep Islands. You see, ISRO has provided telemedicine connectivity to all the five hospitals in the Lakshadweep Islands.'

'Which is the super speciality hospital for them?' I asked.

'It's the Amrita Institute of Medical Sciences, Kochi, AIMS, for short. You see, they have this Indira Gandhi hospital in Kavaratti Island. In October 2003, one P.P. Hameed was brought to the hospital. He is a head constable in Kavaratti police station.'

'What was wrong with Hameed?'

'He was unconscious and sweating beyond normal limits. He had been brought to the casualty section. The doctors there examined him. They couldn't record his pulse. His blood pressure was not readable. His respiratory movements were shallow and weak.'

'What did the doctors do?'

'They took his ECG and found that he had acute heart attack with complete heart block. So they sent the ECG and other details of Hameed to the cardiologist at AIMS through the telemedicine system.'

'What did the specialist advise?'

'He gave the line of treatment to be followed. He also prescribed some drugs. The doctors in Kavaratti followed that line of treatment. They kept in touch with the specialist at AIMS through the telemedicine system and sent him all the details.'

'Did it work?'

'It certainly did. Hameed's condition improved. He soon became conscious and started conversing with his kith and kin.'

'That's wonderful!' I said. 'Do you also use your telemedicine connectivity during disaster situations?'

Satyamurthy cleared his throat. 'Yes, we do. You see, there are these two hospitals in the Andaman & Nicobar Islands, the G.B. Pant Hospital at Port Blair and the Bishop Richardson Hospital at Car Nicobar. ISRO has established telemedicine facilities in these two hospitals.'

'Which is the super speciality hospital for them?' I asked.

'Sri Ramachandra Medical College Hospital in Chennai. At the time of the tsunami in December 2004, this telemedicine network was widely used for disaster relief work for the benefit of the remote population of the Andaman & Nicobar Islands. Let me tell you that something very remarkable happened at that time which showed how robust our telemedicine connectivity was.'

'What was that?'

'You see, at the time of the tsunami, the entire communication network connecting the islands to the mainland broke down. It just didn't work. The only thing that was working was our telemedicine system. So it was used for all kinds of communication purposes.'

'That's remarkable,' I said. 'Tell me something else, do you also have mobile telemedicine applications?'

'Yes. We've tried that and let me tell you that it works beautifully. The mobile applications have generally been in the field of ophthalmology. We have these mobile vans for which we have established telemedicine connectivity. These vans are used for conducting eye camps. Villagers undergo eye screening for cataract, glaucoma and diabetic retinopathy.'

'Where do the specialists come in?' I asked.

'The doctors in the mobile vans keep in touch with the

specialists located in the super speciality eye hospitals through the telemedicine system and consult them in difficult and complicated cases.'

'Which are these super speciality eye hospitals?'

'We've given these mobile telemedicine facilities to Aravind Eye Hospital at Madurai, Shankara Netralaya at Chennai and Prabha Eye Clinic at Bangalore. Their rural eye camps have been very successful. ISRO is now thinking of going in for more numbers of mobile telemedicine units.'

'That's great.'

'One more thing. ISRO has also provided telemedicine connectivity to be used at the time of fairs and festivals. You see, lakhs of pilgrims visit Sabarimala shrine in Kerala every year. We've installed telemedicine system at Pampa, which is at the foothills of Sabarimala shrine.'

'Which is the super speciality hospital?'

'AIMS, Kochi. A large number of pilgrims have availed of the telemedicine facilities and we have been in a position to save some lives, too.'

'It's a good effort.'

'We also have OncoNet that provides telemedicine connectivity to cancer hospitals. It connects the Tata Memorial Cancer Hospital, Mumbai, that is the super speciality hospital to several regional cancer hospitals like the ones at Guwahati, Chiplun and Thiruvananthapuram. The Tata Cancer Hospital has also been connected to several general hospitals in the North-east.'

'Has that helped?'

'Yes, it has been very useful to cancer patients living in remote and inaccessible places to get the benefit of expert consultations. Let me tell you something in general about telemedicine. It has helped in saving cost and effort to the patients in rural areas. After the telemedicine connectivity is provided, the patients in rural areas are not required to travel long distances for obtaining expert consultation and treatment. They save money in the process.'

'Have you made any documentation about the savings in cost?'

'Yes, we have. You see, there is this independent agency which conducted a study of 1000 patients in Chamarajanagar District Hospital in Karnataka.'

'Is that on ISRO's telemedicine network?'

'Yes. It's connected to Dr Devi Shetty's heart hospital in Bangalore. The Narayana Hrudayalaya.'

'Now, tell me, what did the study find?'

'That, for the patients of Chamarajanagar, there was a cost savings of about 81 per cent. What that means is that the patients who availed of the telemedicine consultation and treatment, spent only 19 per cent of the money which they would've otherwise spent if they had to travel to the nearest city for similar treatment.'

'Well, the findings of the study sound quite plausible,' I pointed out.

'Let me tell you another thing. The saving in cost to the patient is much more significant if the patients are from offshore islands and remote, inaccessible places. It also means that the patients not only save on cost, they also get quick and timely medical aid.'

'That makes sense,' I said.

'Well, that's why the thrust of ISRO's telemedicine programme has been to provide connectivity to such remote, inaccessible places. You see, we've already provided connectivity to the Andaman & Nicobar Islands, all the islands in Lakshadweep, several places in Jammu & Kashmir and the North-east.'

'How extensive is your telemedicine coverage in the North-east?'

'Fairly extensive. Our plan is to provide telemedicine connectivity to all the seventy-two district hospitals in the North-east. So far, we have established telemedicine systems in about twenty districts there.'

'For the country as a whole, what is your coverage like?'

'ISRO's network, right now, covers about 144 hospitals all over the country. This consists of 113 remote, rural hospitals and health centres connected to thirty-two super speciality hospitals.'

'That's very extensive,' I pointed out. 'How many patients have been treated by the telemedicine network?'

'More than 50,000 patients so far.'

'That's a magnificent job,' I said. 'Did you hear what some of the patients at the Tripura Sundari Hospital said? That your telemedicine system is a miracle!'

Satyamurthy beamed with pleasure and flayed his rather plump arms around. That must be his way of responding to a well-deserved compliment, I thought. But in the process, he very nearly upset the plastic tray carrying the airline's meagre refreshments.

※※※

Dr Devi Shetty has been a pioneer in popularizing telemedicine. I went across to Narayana Hrudayalaya to talk to him. Narayana Hrudayalaya is housed in a beautiful building, its lobby can easily be mistaken for that of a five-star hotel. The telemedicine room is located at the rear end of the building in the basement.

It was in the telemedicine room that I talked to Dr Devi Shetty. He had just finished having telemedicine consultations with the District Government Hospital, Chamarajanagar.

'You are considered a pioneer in telemedicine,' I told Dr Shetty. 'Tell me, why did you think of adopting it in a big way?'

He gave me a dazzling smile. 'I am a heart doctor. So I can only talk about heart patients. Well, if you want to help heart patients, you have to reach out. But let me tell you that you won't find heart super specialists in most Indian cities. Can I give you some numbers?'

'Yes, of course,' I said.

'Over 65–70 per cent of the medical specialists live in number one cities,' he said with an animated expression. 'Another 20–25 per cent live in number two cities. Rural areas have only 2 per cent of the super specialists.'

'That's sounds pretty dire,' I said.

His forehead creased in thought. 'But the bulk of our people live in rural areas. Most of them can't go to the cities. Let me ask you a question. If somebody has a heart attack in Tripura, can he go to Kolkata for treatment?'

'That would be difficult,' I conceded.

He smiled, his eyes shining. 'See, when somebody has a heart attack, he has only six hours of time. That's called the golden period. In that period, a heart attack has to be diagnosed and thrombolytic therapy has to be given. How do you do that?'

He had a point there, I thought.

'Let me tell you another thing,' he continued. His expressive eyes were grave. 'We need to optimally use the talented workforce. You see, when I was in medical school, my teachers treated only those patients whom they could touch. But now, with the technology available, I'm in a position to treat people I can't even touch because they are thousands of miles away. Since more and more technology is coming our way, we should use it to take away pain and suffering.'

'That's true,' I said.

He said, his eyes resting on me reflectively, 'You see, ideally the doctor should be sitting by the side of the patient, touching, consoling and treating. I can tell you that treating a patient through telemedicine is just short of that. Other than that, you can do everything.'

'Is there any disadvantage of treating patients through telemedicine?' I asked.

His eyes were bright with interest. 'Well, as I told you, nothing is like being there. But if you can't be there, telemedicine is the second best. Because of our institutional access to rural areas, it should be a stimulus for the government and other organizations to create an infrastructure like telemedicine connectivity.'

'Could you give me an example?'

He smiled a seraphic smile. 'Why not? Take the case of Chamarajanagar in Karnataka. It's a tribal area. There is no healthcare in Chamarajanagar. There's a government hospital, but that's like a parking lot for the patients. Nothing happens. So what did we do? We created a CCU. Many patients came and we were able to thrombolyse them.'

'Why, that's fantastic!' I observed. 'What about your partnership with ISRO in providing telemedicine connectivity?'

He beamed me a smile. 'Well, it has been a very productive partnership. You see, ISRO is a great organization. Not only does it rank with the best in the Western countries in making rockets and satellites, but it also has its heart in the right place.'

'Well, coming from a heart doctor,' I pointed out, 'it certainly is a great compliment.'

'I mean it, though,' he said, his eyes bright with excitement. 'ISRO is interested in using its technology for the benefit of the common man. Particularly for people living in remote, inaccessible areas. ISRO and I share this passion. Except I'm interested in reaching medical care to these people. That's why we've forged such a great partnership.'

'Is that why,' I asked, 'you've chosen hospitals in places like Tripura or Chamarajanagar?'

He nodded. 'Absolutely. We have to go to rural areas where they have no hope. There's no point in establishing telemedicine connectivity between Chennai and Mumbai where facilities already exist. We're not interested in second opinion telemedicine.'

'Isn't that how telemedicine is used in the Western countries?' I asked. 'For second opinions?'

He laughed drily, 'That's true. There, they have one cardiologist giving the first opinion and they want a second opinion through telemedicine connectivity. We are not here for that! We want primary healthcare in telemedicine. We want to reach out to common people who have no healthcare at all.'

He then ended the interview as his presence was required for telemedicine consultation with a hospital in Tinsukia in Assam. I took my leave and headed back home.

On my way back from Narayana Hrudayalaya, I played and replayed Dr Devi Shetty's words in my head.

'ISRO has its heart in the right place. It cares for the common man and his plight.'

I smiled. The patients in Udaipur were right. The good doctor was dead right in his diagnosis all the time, it seemed.

6

IN ORISSA
Helping Flood Efforts

Devi has always been a wild river. The story goes that Devi was in love with the river Kathjhori. They lived together in harmony for a while. Then, one rainy day, they had a violent quarrel. In a fit of rage, Devi flew away all the way to the Bay of Bengal. But the rage did not abate; every rainy season, Devi vents its rage on the hapless villages that populate its banks by sending down the most devastating floods. The story maybe apocryphal. But floods do come in the river Devi every year like clockwork. When they peak, lives are lost and properties are damaged. Devi breaches its embankments, submerges vast extents of agricultural lands and decimates communication networks, leaving large numbers of human settlements marooned.

The worst ever flood that Devi sent down in the last century was in 1955. One rainy night, the bloated Devi breached its embankment at Dalei Ghai. Floodwater gushed through the rapidly widening breach. The torrents washed away large numbers of villages, people and cattle. The villages, that were not washed away, remained marooned for days on end. The road from Cuttack to Jagatsinghpur, the lifeline of the entire region, was cut-off in several places.

Telephone and telegraph links snapped, because the poles carrying them were twisted by the flood torrents. Trees were uprooted. Large tracts of land were under inundation for weeks together. Relief could not be delivered quickly because of the total breakdown of the communication network. Boats were pressed into service to ferry relief to the affected, but they could not negotiate the terrain because the severed trunks of coconut and palm trees were floating around aimlessly, buoyed by the eddying waters.

Water finally receded after what seemed like eternity, but the devastation was complete. The telltale marks of Devi's fury could be seen long after the floodwater had receded. Silt, thick as treacle, lay almost everywhere. Heaps of sand, shaped like dunes in a desert, stood for years as testament to Devi's devastation. Damage to the crops, land, houses, property, public utilities, and road and telecommunication networks was considerable. The total extent of damage was estimated at Rs 12.11 crore; the highest ever flood damage in the history of Orissa till that time. The way Devi vented her rage soon became the staple of Oriya folklore, but it also influenced policy decisions in the government of Orissa. After the breach in Dalei Ghai, the water level of 1955 officially became the determining factor for stipulating the plinth of the dwelling units.

Devi again breached its embankment in 1982 and ravaged the countryside. But this time around, other rivers in Orissa's deltaic region went wild too. The devastation they caused together was severe. Fifty-four lakh people in eight districts were rendered homeless; twelve lakh hectares of cultivated land were submerged; five lakh houses were either submerged or washed away; and 127 human lives were lost. The livestock casualties were of the order of 26,359. All in all, it was severe flood by any standard.

Severe floods have now become regular visitors to the deltaic plains of Orissa. Since 1965, Orissa has experienced floods for seventeen years. Why does this happen? One has to put it in historical perspective. The deltaic region in Orissa was transformed, over a 125-year (1803–1928) period from a flood-dependent agrarian region to a flood-vulnerable landscape. The transformation came because of the changes made by the British. They put in place a new regime of property rights; they also made a number of technical interventions. Over a period of time, there was a movement from embankments to a canal system.

After the British left, a definitive change came in the shape of Hirakud dam built across the river Mahanadi. It was to moderate the floods in the deltaic plains of Orissa that the dam was built. It is an abiding irony that the frequency of floods has increased after the construction of the dam. The facts speak for themselves: the average difference between the floods now is 3.75 years as compared to 9.75 years earlier.

Why is it so? There are several reasons. For one, because of massive deforestation, the increased run-off from the upstream catchments has made the storage capacity of Hirakud dam inadequate. In any case, the original storage capacity of the dam was designed on the basis of the past trends in the run-off from the upstream catchments; trends which are no longer valid. As if to compound it, there is the problem of siltation. This has reduced the original capacity of live storage of the reservoir. This is because soil erosion in the upstream catchments has occurred at a rate higher than anticipated originally.

The other reason is purely conceptual. The idea of controlling floods is in conflict with other objectives why a dam is built, such as irrigation and power generation. To meet the last two objectives, the reservoir needs to be filled up very quickly and the level of the reservoir has to be retained at the highest possible level. So the reservoir cannot hold floodwater from the returning monsoon or cloudbursts. That being the case, water has to be discharged in bulk from the reservoir and this leads to floods downstream.

The question still remains: Why is deltaic Orissa so flood-prone? One reason is that the existing natural drainage is not able to carry the floodwater within its course to the sea. That is why the floodwater overflows the banks of the rivers and causes breaches in the embankments. Other reasons are inadequate embankment support, absence of flood plain zoning and haphazard development in flood plain, lack of efforts at river treatment and diversions, and storm surges because of cyclones.

To add to Orissa's woes, there are intriguing changes in its ecosystem. Out of the traditional six seasons, Hemanta—the season that follows autumn—has totally disappeared. Some other seasons now last merely three or four days. The birds in Orissa have started behaving strangely, too. The Haladi Basanta—the black headed oriole—does not herald the arrival of spring as it used to do before. Even the trees now behave oddly. Mango trees used to flower in November while the mahua flowered in February. Now the mango trees flower in September while the mahua tree flowers in December; that is, a full two months ahead of what it used to be.

Perhaps, Orissa's vulnerability to floods has something to do with how it is located geographically. It sits at the top of the Bay of Bengal where the weather is formed. So even a slight change in Bay of Bengal's behaviour has an impact on coastal Orissa. The bay becomes the centre of low pressure causing heavy rains in Orissa and bringing down severe floods.

※※※

What has ISRO done to help Orissa's flood efforts? I happened to be in Bhubaneshwar in the last week of June 2001, and I walked across to Behera's office to find out. Behera was poring over a number of brightly coloured satellite imageries when I caught up with him.

'What are those imageries for?' I asked him.

Behera squinted. 'Oh, we're doing some work on watershed prioritization.'

'I came here on a different mission,' I explained. 'To find out what ISRO has done to help Orissa's flood efforts.'

Behera brightened up. 'We have done a lot,' he said exultantly. 'ISRO has provided important inputs on some vital parameters of flood forecasting.'

'What kind of parameters?'

Behera stroked his chin thoughtfully. 'Parameters such as catchment characteristics, soil moisture, hydraulics and hydrology,' he explained. 'You see, ISRO's remote sensing satellites do provide valuable information on these vital parameters.'

'What about ISRO's communication satellites?'

'They provide very important inputs such as event capture at the time of floods. You see, ISRO's communication satellites carry Charge Coupled Device and Advanced Very High Resolution Radiometer cameras that have a spatial resolution of 1 kilometre. These are used for real time coarse observations of the floods.'

'All right,' I said. 'But those are only indirect inputs. What do you do directly to help?'

'We help in several ways,' Behera said earnestly. 'For one, we've identified erosion-prone areas, that is, those vulnerable areas along the river-banks which are subject to erosion.'

'How do you do that?'

'By using geomorphological maps of the delta area. We've identified points on river embankments that are vulnerable to breaching. We maintain the information in a data bank that is stored in a geospatial framework.'

'What do you do with the information?'

'We pass it on to the district authorities, so that arrangements can be made to watch the vulnerable points constantly. The local villagers guard them on a round-the-clock basis. That is, when the rivers are in their peak flow.'

'What do the local villagers do really?'

'Well, the district administration provides sandbags to such locations in advance. The local villagers use the sandbags to plug water holes. And also, to check over-spilling of the embankments.'

'What other things do you have in your database?'

'You see, we had the Super Cyclone here in 1999. After that, a large number of multi-purpose cyclone shelters have been constructed in coastal Orissa. The database that we have includes the location of these cyclone shelters. Now, when people are evacuated from their homes, they go and stay in these cyclone shelters.'

'Do you have other things in your database,' I asked, 'things that can directly help in relief and rehabilitation efforts?'

'Lots of things. The database has information in respect of boats and boatmen. This is the kind of information that is used for rescue and relief operations.'

'That's useful.'

'The database also has information on earthmovers. You see, earthmovers are used for road clearing. This is a priority item because the roads get cluttered up with debris at the time of floods. It has information on godowns or places where food grains and other relief materials can be stocked. It also has information on hospitals or health units stocking medicines and providing medical care. You see, this is important because sometimes epidemics break out at the time of floods.'

'That's true.'

'Our database also has information on vehicles like trucks, tractors or jeeps and the places where they are available. You see, these vehicles are used at the time of flood for transporting relief material as well as for transporting staff deployed for relief and rescue operations.'

'Who exactly do you give this information to?' I asked.

'To all the user agencies. Particularly to the district administration. Let me tell you that there is one service that ISRO provides which is unique and helps a great deal in the flood efforts.'

'What's that?'

'ISRO provides timely information about the area inundated by the floods. We provide such information to the user departments.'

'How does it help?'

'It helps the user departments in organizing relief operations. You see, the most critical element in monitoring and management of floods is the timeliness of information.'

'That's true.' I said. 'How quickly do you provide this information?'

'In about six hours.'

'That's pretty quick.'

'ISRO also generates flood damage information. For that, we use the satellite data acquired during flood, pre-flood and post-flood times along with ground information. On the basis of that, we estimate the district-wise flood inundated area, the crop area inundated, the number of villages marooned and the communication network affected.'

'Tell me, what are the things you factor in while estimating the flood damage?'

'Several things, such as duration of the flooding, magnitude of the flood, the number of flood waves, area affected and the type of land use features.'

'What is this damage assessment used for?'

'Basically for the government and NGOs to provide relief. These relief measures can be both short term and long term. The short-term measures enable those affected by the flood to resume their normal lives as soon as possible. The long-term measures are to help the community to cope with the aftermath of the flood.'

'That's a laudable effort,' I said.

✳✳✳

Behera telephoned me the day before I was due to leave for Bangalore.

'There have been heavy rains in the catchment area,' he said. 'There is a strong likelihood of the coastal districts being flooded. Why don't you stay back for a few days and see how ISRO assists the administration in their flood efforts?'

'I'll stay back,' I readily agreed.

I started unpacking my bag for an extended stay.

✳✳✳

Behera rang me up the next morning to announce, 'Preparations have already started in ISRO for preparing the maps. They worked the whole of last night, preparing the satellite coverage charts over Orissa, based on the data from the remote sensing satellites. They'll be sending the maps to me today.'

'What'll you do with the maps?'

'Right now, I'm making arrangements here to send the maps electronically to the relief agencies. I'll also generate hard copies of the maps. And I'll distribute the hard copies to the relief agencies also. Why don't you come over and see how we do it?'

I went over to Behera's office. The whole place was a beehive of activity. The maps had been received electronically and downloaded. They were being sent electronically to different departments of the government and the district administration. Hard copies were being made and dispatched to different destinations.

'Have you heard the latest?' Behera asked.

'No,' I said. 'I've been confined to my room in the Circuit House. You see, it has been raining all the time.'

'Well, it has been raining even harder in the catchments of all the rivers. The reservoir levels are up. I've heard that the authorities had no option but to release water from Hirakud reservoir.'

'So what's going to happen?'

Behera grimaced. 'It'll send down horrific floods. Lakhs of people in coastal Orissa will be affected. The situation is grim.'

'What are you going to do?'

'I have to ask for fresh inundation maps to be prepared.'

The task was cut out for Behera, I thought.

※※※

Next morning, I read in the newspaper that water had been released from the Hirakud reservoir. And the whole of coastal Orissa had experienced unprecedented floods.

I telephoned Behera. He took some time to come to the telephone. When I finally got him on the line, he sounded distressed.

'There have been severe floods in coastal Orissa,' he said. 'This morning, I got the inundation maps. I've sent them off to the relief authorities.'

'Is the situation very bad?'

'Yes.' His voice was grim. 'I believe our maps have been very useful. There are people guarding the embankments of rivers with sandbags.'

'I've a favour to seek,' I said. 'I'd like to see how the relief operations are actually going on. Do you think you can talk to some district collector who could possibly arrange for me to visit a village and see how things are?'

There was silence for a minute or so. Was I asking for too much?

Behera said, 'People in the district administration will be too busy. But I'll try. There is this chap who is the collector of Jagatsinghpur. I know him rather well. I'll talk to him. Can I come back to you later during the day?'

When Behera talked to me a few hours later, his voice was cheerful. He said, 'I managed to get through to the collector, Jagatsinghpur. I told him about your request. It seems you are in luck.'

'Why, what happened?'

'The collector is going tomorrow by launch up the river Devi to review relief operations at a place called Devi Deula. He has agreed to take you with him.'

'Thanks a million,' I said.

'Well, I'll make arrangements for you to go to Machhgaon by road. That's the place where the collector's launch is parked. But you have to start for Machhgaon by the crack of dawn.'

'I'll be ready,' I said and rang off.

※※※

Machhgaon is the place where Devi meets the Bay of Bengal. The collector was late by an hour. Poor chap, I thought, he must be a very harassed man! A million things to do, I suppose.

The water was at low ebb and the collector's launch was moored at the far end of the mudspit. There was nothing remotely pretty about the vessel as it sat awkwardly in the water. It was badly dented in several places and its hull had the battered appearance of scrap metal. But once we ascended it, it turned out to be rather large and commodious. The lower deck was a wide space, curtained with sheets of tarpaulin and with several wooden benches. The galley and the engine room were at the opposite ends of the lower deck. On top, there was an equally commodious upper deck, with a wheelhouse and large-sized cabin. There were two deep, cushioned chairs near the wheelhouse, in the shade of a canvas awning.

The launch pulled away from Machhgaon, with its engine alternately spluttering and hammering. It now picked up speed and chugged up the river. From the launch deck, I looked at the river Devi. It lay in spate, just a couple of feet beneath its bank, swollen with raging water and flotsam. The current was very swift and the river had become a foaming sheet, carrying away heaps of water hyacinth, rotting timber and all kinds of debris. There was something forbidding about the river Devi, I thought.

After what seemed like eternity, we reached Devi Deula. The launch stopped a little distance away from the river embankment. There were wooden, country boats waiting in the water to transport us to the river embankment. The collector went off to supervise relief operations at the sub-registrar's office and I walked along the embankment. The water level in the Devi was perilously high; just about 2 feet below from where I stood on the riverbank. Suddenly, there was a gust of wind and the eddying waters of Devi hit the embankment with a mighty swirl. The water rose up in a vicious spray and I got the full blast of the muddy water on my face. It was the feisty Devi's way of reminding me that she was not predictable in the least, and always to be watched out for.

Sandbags lay everywhere, piled up high on the embankment. There were groups of people putting sandbags below the embankment, just above the water level. I walked across to a group.

'Can I talk to you?' I shouted against the wind. I could feel the wind carrying my words away to the villages below the embankment.

But they heard me. They stopped putting the sandbags and climbed up. Soon the group had gathered around me.

'Where are you from?' I asked the tall man who stood next to me. His dhoti, caked with mud, was hitched up high above his knees and that's all he wore. Thick stubble sat uneasily on his angular face; quite a few days' growth, I surmised. He lit a bidi before answering me, took a few puffs and exhaled the smoke noisily.

'From Gajrajpur,' he said finally, and gestured towards a village below the embankment. 'I am Jaga Jena.'

'Why are you putting these sandbags?' I asked.

Jaga Jena narrowed his eyes. 'We're putting these bags in places in the embankment where there are water holes,' he said, 'so that Devi doesn't breach there.'

'But who tells you where it'll breach?' I asked.

He inhaled deeply on his bidi. 'The RI,' he said laconically.

'The RI?' I asked. 'Who's he?'

It was another person in the group who answered my question. 'The revenue inspector,' he explained. 'He is from the revenue department.' He introduced himself now. 'I am Gobind Acharya. I am from Bhatpada.' He pointed to a village just by the side of Gajrajpur.

I looked at Gobind Acharya. He was better dressed than the others. And more articulate.

'What does the RI tell you?' I asked him.

Gobind Acharya's voice had a priestly rap. 'The RI has a map, which shows the weak areas in the embankment where there have been some erosion in the past. Those are the places where the embankment is likely to breach.'

'And also, where there are water holes.' That was a new voice. He introduced himself. 'I am Baikunth Padhi from Devi Deula.' He pointed to his village. It was just below the embankment. I could see a temple spire rising above the coconut trees.

'Do you people keep on putting these sandbags all the time?' I asked.

'Yes, yes,' Gobind Acharya said. He seemed to be the leader. 'We keep a watch round-the-clock. We have to be careful, you know, but it is only at times like this when the water level in Devi is very high.'

'Do you do this at night also?' I asked.

'Yes,' Gobind Acharya explained. 'We light petromax lamps and keep vigil.'

Jaga Jena chipped in. 'One has to be careful, particularly when it is Devi. She is kind of wild, you know. She always attacks at night.'

He talked as if Devi was a predator, like a tiger waiting to pounce in stealth.

'Who decides whose turn it is to keep vigil?' I asked.

'The villagers sit down and decide,' Gobind Acharya said, gesturing to the villages lying under the embankment of the river. 'We've prepared a chart showing who should be on duty, what they should do, what points they should guard. A kind of duty chart for all the able-bodied people of these villages.'

'Who provides the sandbags?' I asked.

'The revenue department,' Baikunth Padhi said. 'Whenever we need more, we just have to tell the RI. In any case, there is plenty stocked near the sub-registrar's office in Devi Deula.'

'Tell me,' I asked, 'what about other relief materials? Things like food grains?'

'All that have been taken care of,' Gobind Acharya said, wringing his hands. His tone was smug. 'Stocked up well in advance.'

'Where are they stocked?' I asked.

'In the cyclone shelter,' Baikunth Padhi explained. 'You know about that Super Cyclone in 1999, don't you? After that, the cyclone shelter was built. It's in an elevated place. The floodwater, however high, can't reach the cyclone shelter.'

'That's where we'll all go when the floodwater enters our houses,' Jaga Jena said, running his hand over his stubble.

It was Gobind Acharya's turn to intervene. 'That is when evacuation is ordered by the authorities,' he said, waving a genial finger in Jaga Jena's face. 'That's what the collector is discussing right now with all the local officials in the sub-registrar's office.'

'What a pity that people didn't do these things when Dalei Ghai broke!'

The words came from a man who was sitting a little away from us. He looked quite old: he had a wizened face with deep, crinkly lines and a mop of hair that was silvery white.

'Who are you?' I asked. 'Where are you from?'

He took his time to answer. 'I am Gopal Bahalia,' he said haltingly. 'I am from Bahalia Sahi.'

'Where is that?' I asked.

It was Jaga Jena who answered, scratching his stubble, 'Just next to Gajrajpur.'

Gopal Bahalia now took out his batua, the bag that stores the accessories for making a paan. The bag was patched with brightly coloured pieces of cloth. Just like the ornamental appliqués in a canopy.

'You said something about Dalei Ghai, didn't you?' I asked Gopal Bahalia.

'Yes,' he said, with a frown spreading across his wizened face. 'You see, that time, they didn't put sandbags, because they didn't know where to put them.'

'You mean to say,' I asked, 'that the Devi wouldn't have breached that time if they had put sandbags at the vulnerable places?'

He did not answer immediately. Instead, he pulled the

drawstring to open the batua and took out a paan leaf, a betel nut, some lime and a nutcracker shaped like an elephant's head. He cracked the betel nut, the sounds ringing out sharp and crisp like a woodpecker tapping tree trunks in search of insects.

He said at last, 'Let me tell you that I went there a few months after the breach. The villagers told me that it wouldn't have happened if they had guarded the vulnerable points in the embankment day and night and put sandbags there.'

'What would have happened if they had put sandbags on the vulnerable points?' I asked.

Once again, he did not answer immediately. He now cut the betel nut into tiny pieces; as tiny as a cat's tooth. He spread lime on the paan leaf, making white patterns on the bright green leaf, like the stripes on a squirrel's back.

'You see, in that case all the water would have gone to the sea in Devi itself,' he said with a crowing laugh.

He now gathered the betel pieces to the paan leaf in an expert, unhurried motion. He rolled the leaf into a cone and secured it at the top with a clove. He opened his mouth wide, put the paan in and started chewing it with relish.

'How old were you when the Dalei Ghai breached?' I asked.

'Ten years old,' he said. He pointed to all others in a derisive gesture. 'And all these fellows were not even born!'

'That's true,' Gobind Acharya conceded with a dry laugh.

'What happened when Devi breached at Dalei Ghai that time?' I asked Gopal Bahalia.

'All hell broke loose,' he said, still chewing on his paan. 'The water came gushing early in the morning. How much water! Torrents and torrents of it. Drowned everything in sight. We put the women and children on the roofs of the houses.'

'What did you do?' I asked.

He laughed aloud and little bits of the paan came flying out of his mouth as if in a shower. 'I climbed a coconut tree. But tell me, how long can one stay on a coconut tree? Or, on a thatched roof?'

'It must have been terrible.' I said with a shudder.

'Yes, it was,' he said, spitting out the paan juice. 'Do you know something? Relief didn't come for days. Finally, the boats came. But it was too late!'

'What kind of relief?' I asked.

He continued as if he had not heard my question. 'As if God hadn't punished us enough, it happened again.'

'When was that?' I asked.

'A few years later. Devi breached again at Ali Pingal.'

'Did it?' I asked. 'Where is Ali Pingal?'

'Near Jagatsinghpur,' Gobind Acharya intervened to say. He seemed to know the geography of the place rather well. 'A few miles upstream from here.'

'Yes, at Ali Pingal,' Gopal Bahalia said emphatically. 'And believe me, it was the same thing like at Dalei Ghai. People didn't guard the embankment. They should have put sandbags in the vulnerable points!'

'Because nobody told them,' Gobind Acharya said sagely. 'In any case, they wouldn't have known where the vulnerable points were.'

Gopal Bahalia spat out some more of the paan juice. 'You can say whatever you like! But only bad things have happened. Devi has punished us in the past.'

'What about this year?' I asked.

He brightened up. 'Things are good. See how much water there is in the Devi! Much more than other times. But these boys here are guarding the embankment day and night. Putting sandbags at all the weak spots. Relief materials are ready and stacked away. Things certainly look good.' He nodded approvingly.

A figure in white uniform approached us. He gestured to me and said, 'The collector saheb is calling you, sir. We are ready to go back.'

I took leave of Gobind Acharya, Jaga Jena, Baikunth Padhi and Gopal Bahalia.

※※※

The collector's launch began its return journey. Its engine spluttered and hammered alternately and the launch started chugging. From my position on the open deck, I looked at the river in spate: Devi was certainly a terrifying sight.

I sensed someone joining me on the deck. I turned around to see. It was the collector. We stood side by side and surveyed the river.

'So you work in ISRO,' the collector asked.

I nodded. 'Yes, I look after their finances. Tell me something. Are ISRO's inputs useful in your flood efforts?'

'Yes,' he said, 'in several ways. You see, ISRO has given us inputs on erosion-prone areas. These are very valuable inputs in the sense that, they tell us which are the vulnerable areas along the river embankment.'

'Yes,' I said, 'I saw that for myself. People keeping vigil and putting sandbags on water holes in places which are indicated.'

The collector said, waving a hand in the air, 'ISRO helps us in yet another way. It provides us with inundation maps. That helps us in organizing relief operations.'

'But do you get them on time?' I asked.

He nodded. 'Yes, we do get them on time. ISRO sends them across to us on email. Let me tell you that I got the last one within six hours of the inundation.'

'What about estimation of the damage?' I asked.

'Yes, that too,' he said. 'ISRO gives us very valuable inputs in estimating damage. They do it on the basis of their satellite imageries. But they also fly their aircrafts at a very low height and photograph the areas affected. They pass on the damage information to us.'

'Does that help?'

'Immensely. You see, providing compensation to the flood victims has always been a very contentious issue. The flood victims often complain they haven't got enough. With ISRO's inundation maps, we get objective data. The satellite imageries can't lie, can they?'

'No, they can't,' I readily agreed.

'There is something else,' he said reflectively. 'We had this Super Cyclone in 1999. At that time, I was the collector of another coastal district which was totally ravaged by the cyclone. ISRO was very helpful at that time.'

'Why, what did it do?'

'It gave us advance warning of the cyclone.'

'Did it? How many days in advance?'

'At least two days before the cyclone. We should've evacuated the people. But you know how difficult it is!'

We heard someone coming up the steps. He was the collector's deputy carrying a sheaf of papers and he wanted the collector to see them urgently. The collector went off, leaving me all alone to survey Devi in its unremitting rage.

※※※

It was time for me to go back to Bangalore. I went to Behera's office to take leave of him and thank him for all his help. I found Behera poring over maps, occasionally barking out orders to his people to send out hard copies to different government offices.

'So what's happening now?' I asked.

'People in ISRO are still collecting satellite data to monitor the flood situation in Orissa,' Behera explained. 'They are sending over the inundation maps.'

'And you're sending them out to the user departments. How long will you continue doing it?'

There was a determined expression on Behera's face. 'As long as the flood lasts. Till we observe from the satellite imageries that the flood inundation has started receding. And there's no fresh inundation.'

'What will you do after that?' I asked.

He smiled. 'Oh, well! Then it is time to prepare the flood-damage information'.

Floods may cease but ISRO's work goes on, I thought.

7

KARNATAKA'S PANCHAYATS
Empowering Grass-roots Representatives

Girijavva Ganigar had a big dream. An illiterate, backward-caste woman from Tada village in Karnataka, she wanted to be elected to her panchayat as a member. But like the obstacle in the best stories, there were problems in her case, too. For one, she was a woman and the society does not take very kindly to women in representative roles. Even people from her own caste were opposed to her being in the panchayat. Undaunted, Girijavva contested the elections and won her seat. She was even elected as the vice-president of the Tada panchayat.

Then one day, one of the members of the Tada panchayat who belonged to the upper caste, refused to attend the meetings of the panchayat because a Dalit woman was elected as its vice-president. He refused to pay the water tax he owed to the panchayat. The panchayat then issued him a notice. One day, he stormed into the room where the panchayat meeting was in progress and snatched away the attendance register. He signed it only to indicate that he was present in the previous meetings of the panchayat which he, in fact, did not attend. To add insult to the injury, he abused both the president and vice-president in choicest invectives for daring to send him a notice.

Girijavva was appalled by such outrageous behaviour. She did the unthinkable: she telephoned the BDO and asked that he should immediately intervene to discipline the errant member. The BDO did intervene and set matters right. As a result of Girijavva's initiative, people of Tada village hold her in great esteem.

The fact that an illiterate woman like Girijavva is aware of her role and responsibilities as a member of the panchayat was made possible only because she had participated in an intensive training programme conducted for women representatives of panchayats. The training programme had two objectives. One, to enable them to function confidently and assertively as members of the panchayat. Second, to familiarize them with the functioning of the panchayats so that they could effectively exercise their role in the political system of local self-governance.

※※※

Is there a need for such training? The question has to be viewed in the context of the Seventy-third Amendment to the Constitution of India and the legislations that were passed by the state governments providing for reservation of women in panchayat bodies. For example, the Karnataka Panchayati Raj Act, 1993, provides for one-third reservation for women in respect of seats and authority positions in all panchayati raj institutions. As a result, a large number of women have been elected to the village panchayats; in fact, as against the stipulated one-third, 45 per cent of the seats were won by women.

It is heartening to see such a large number of women being elected to the panchayats. But there is a problem: a sizeable proportion of these women are first-time/first-generation representatives. That is why questions are asked about the capacity of these women representatives to deliver in the political space that is so alien to them. It is also not a sheer coincidence that very few elected women representatives at the panchayat level are familiar with the legal provisions, the structure of the system, the rules and regulations that govern panchayat bodies, or with even their own rights and responsibilities as members. This is why providing training to these women representatives is important.

In February 1995, over 600 women members of panchayats in Karnataka participated in an exciting training programme, using satellite-based interactive communication system. Groups of thirty-odd women representatives assembled in the District Training Institutes of each district to view programmes on a TV set, pose questions to a panel of experts in the studio and hear each other's questions and the panel members' answers through the TV.

The system ISRO used for the training programme was the two-way-audio, one-way-video teleconferencing system. The network consisted of a teaching end (studio) that originated the training material and uplinked the television signals to an INSAT satellite by means of an uplink Earth station located at the studio. The INSAT satellite received, amplified, frequency translated and transmitted the signal over the coverage area. The signals were received at each District Training Institute by a low-cost Direct Reception System consisting of a 8–12 feet antenna, low noise block converter, tuneable receiver and an ordinary TV set.

The transmission over the satellite was for three hours every day. The telecast consisted of a dramatized history of the panchayati raj idea, womens' accounts of how they became panchayat members and their experiences, plays describing the rules, procedures and structures of the panchayat and depiction of ways in which the women representatives are cheated of their rightful place in the panchayat structure and group discussions on issues such as reservation, gender differences in approaches to functioning as panchayat members, obstacles created by families and local communities, and caste identity as a potentially divisive factor.

For the rest of the day, there were resource persons

drawn largely from NGOs who facilitated discussions among the trainees, helped them to analyse their own experiences in the light of what they had seen in the telecasts and clarified issues and doubts. The highlight of the training programme came on the last day of the transmission. It was a lively play presenting the anti-arrack agitation launched by rural women in Andhra Pradesh as an example of collective action for the common good. After the play, there was a talkback session with participation of women from different districts.

On all accounts, the training programme was a big hit with the participants. The most popular part of the telecast was the play on the anti-arrack agitation, with trainees from different districts calling in to voice their support for the action by the women in Andhra Pradesh. The trainees were very happy to see and hear rural women like themselves in interviews and group discussions. They were particularly enthused by the drama sequences, because they were entertaining and helped them to see familiar situations in a new light.

All in all, the satellite-based interactive training programme for the women members of the panchayats was a huge success.

※※※

The other satellite-based interactive training programme for gram panchayat members that was taken up in Karnataka in five phases lasting from November 2002 to July 2003 was more ambitious in scope. It covered 18,000 elected members of 1310 panchayats in forty-four talukas. It also had, as its major target, the women representatives and those from the disadvantaged groups of the society. The only difference was that the receiving centres of satellite communication were located in the taluka panchayats as opposed to the district headquarters in 1995.

The training programme was divided into five phases, each of two days' duration. For the first four phases, the programme consisted of viewing films based on the experience of elected gram panchayat members. The format of the films included an artistic mix of workshop discussions, individual interviews and plays.

Phase five was non-film based: it used theatre techniques for communication, leadership and self-development. The overall focus was to enhance the confidence and self-esteem of the panchayat members—particularly the women members—for their day-to-day functioning, including for example, standing before an audience to deliver a speech or making an intervention during the deliberations of the gram panchayat or Gram Sabha.

The programmes were transmitted from the Satellite Communication Centre located at Abdul Nazir Sab State Institute of Rural Development, Mysore, and were received at the district/taluka receiving stations. The system used was the one-way-video and two-way-audio system. It consisted of a studio located at Mysore, from which the anchors and panelists facilitated the training programme through live or pre-recorded presentations, discussions, demonstrations and talkback sessions. These were transmitted by an ISRO communication satellite through an Earth station, which was linked to the studio at Mysore. The satellite relayed the TV signals for reception directly by VSAT terminals and relayed through TV monitors located at the training centres in different talukas in the state.

The participants sitting at the training centres in the talukas raised their doubts and sought clarifications from the resource persons at the studio in Mysore on an audio channel or fax through telecommunication lines located at the training centres. At the studio in Mysore, the questions received from the training centres were looped back on the audio channel of the TV signal emanating from the studio, so that the questions asked could be heard at all the training centres in the different talukas. The response to the questions went on the TV signal and was received by all the training centres.

✳✳✳

I travelled to Binkadakatte village in Gadag district to find out what had been the impact of the satellite-based interactive training programme. The members of the panchayat were present in full strength to share their experience with me.

I was introduced to the members present. They were Mrs Gowravva Umesh Shivanappanavar, Mrs Dilshad Begum Davalasab Challamada, Mrs Sharada Bhimareddy Moolimane, Mrs Lakshmavaa Hemareddy Koli, Shri Venkappa Yallappa Makali, Shri Anilkumar Somareddy Tirlapur, Shri Tirakappa Nagappa Rangappanavar, and Shri Mallappa Neelappa Karur.

The president made a small speech, welcoming me to their midst and thanking me for having taken the trouble of coming all the way from Bangalore to talk to them. After the speech, I started asking my questions.

'Did you participate in the satellite-based interactive training programme?' I asked.

'Most of us did,' they said in a chorus.

'Did you like it?' I asked.

It was Mrs Shivanappanavar who replied, 'Yes, we did. That was because it was done so differently. They showed us so many films! You know something? We could even ask questions. Or clear our doubts.'

'What were these films about?' I asked.

'On several topics,' Mrs Sharada Bhimareddy Moolimane explained. 'Like the history of the panchayati raj institutions, reservation principles, gram sabha, rules about gram panchayat meetings.'

'You left out some topics, Sharada,' Mrs Lakshmavva Hemareddy Koli pointed out. 'There was also a film on proxy elections and proxy participation in panchayat meetings.'

Mrs Moolimane said rather apologetically, 'Yes, yes. I forgot. I now remember there were films on other topics also. Like education, health, childcare and water supply schemes.'

'You people forgot the best film that was shown,' Mrs Dilshad Begum pointed out. 'It was called "In Sitamma's Footsteps".'

'What was it about?' I asked her.

'It was a wonderful film,' she said. 'It was about the anti-arrack movement in Andhra Pradesh. How it grew out of the literacy programme in Nellore district.'

'But what was the relevance of the film for the gram panchayat?' I asked.

'Well, it was very relevant,' Mrs Dilshad Begum was quick to point out. 'You see, the film raised the issue whether gram panchayats were meant merely for implementing schemes designed elsewhere or whether they had a role in the overall development of the people in the panchayat.'

I looked at Mrs Dilshad Begum with approval. She put the whole thing rather nicely and it was unusual for a woman member in a far-flung village panchayat to do that.

'There was this film on the common property resources of the village,' Venkappa Yallappa Makali said. 'The common property resources like the Gomalas and tanks. How they have to be maintained by the gram panchayats. I liked that film very much.'

'Yes, yes. That was a good film,' Anilkumar Somareddy Tirlapur pointed out. 'It taught us that if the gram panchayat neglects to look after the Gomalas or tanks, the entire village would have to pay a big price. But let me tell you something. There was another film which I liked very much.'

'What was the film?' I asked.

'It was called "Dreaming Together",' Tirlapur said. 'It was about the hopes and aspirations of the elected members of the gram panchayat. What they would like to do for the development of their gram panchayat during their tenure.'

'I am happy you people remember the details of all these films,' I pointed out. 'I wouldn't have remembered even half of them! Were there discussions in the training programme?'

'Plenty of discussions,' Tirakappa Nagappa Rangappanavar said. 'Do you know what happened? We narrated our experiences. We asked questions whenever we had doubts. And we got answers, too.'

'What do you think?' I asked Mrs Dilshad Begum. 'What was the best part of the training programme?'

She was thoughtful for a minute. 'I think the best part was the knowledge that the training programme gave us about our roles and responsibilities as panchayat members,' she said. 'What we are supposed to do to become model members.'

'Dilshad is absolutely right,' Mallappa Neelappa Karur said. 'I would go even a step further. The training programme taught us about the powers of the panchayats, about their responsibilities. The things we learnt at the training programme have made us much better equipped to do our jobs as elected members of the panchayat.'

'That's very creditable,' I said. 'But do you think all that knowledge you gained in the training programme has really helped you?'

It was Mrs Dilshad Begum who spoke now. 'Of course it has helped. You see, with the knowledge we gained, our panchayat has achieved a big victory. Let Trilapur tell you about it. Tirlapur, why don't tell him about our panchayat's experience with the PDS?'

'What is this PDS?' I asked.

'Public Distribution System,' Trilapur said. 'That's the system under which government food grains and kerosene are distributed to the people who are poor. They are given ration cards under the PDS to buy food grains and kerosene at a concession. There were complaints that cardholders were not getting the commodities in time and in prescribed quantities.'

'There were also complaints of corruption,' Malappa Karur pointed out.

'Who's in charge of distribution under the PDS?' I asked.

'It's the food department,' Tirlapur said. 'It comes under the tahsildar, the district food officer and the deputy commissioner.'

'Let me tell you what we had learnt in the training programme,' Mrs Shivanappanavar said. 'We learnt about the powers and responsibilities of the gram panchayat. We were also told about the provisions of the Karnataka Panchayati Raj Act in our training programme.'

'What does the Act say?' I asked.

Mrs Shivanappanavar explained, 'That it is the panchayat which should distribute the ration cards and take the responsibility of visiting ration shops, checking the stock and actually distributing food grains.'

'So what did you do?' I asked.

'We sat down in the gram panchayat meeting and discussed the whole thing,' Tirlapur said. 'We decided that henceforth the distribution of food grains and kerosene should be under the supervision of our gram panchayat.'

'That's very nice,' I said.

'Well, we went and met the tahsildar and the district food officer,' Tirlapur continued. 'They agreed that the panchayat could do it. They also decided that in order to prevent leakages, there should be double-locking of the Fair Price Shop. The gram panchayat should put the second lock.'

'Looks like your mission was accomplished, wasn't it?' I said.

'Not quite,' Tirlapur said. 'Things didn't end there. The deputy commissioner, Gadag, acted now. And let me tell you that he had not lifted his little finger to plug the leakages in the PDS at any time!'

'What exactly did the deputy commissioner do?' I asked.

'The deputy commissioner stayed the order of the district food officer,' Tirlapur explained. 'On the ground that he was not sure of the law on the powers of the gram panchayat in respect of the PDS.'

'What did you people do?' I asked.

'We didn't give up,' Tirlapur said. 'We took the matter to the government of Karnataka in Bangalore. We believe there was a lot of discussion and debate. But we won finally.'

'Why, what happened?' I asked.

'Because the government agreed with us,' Tirlapur explained. 'So now we have the double-locking system. We now supervise the distribution of food grains and kerosene under the PDS. Let me tell you that ours is the first gram panchayat in Karnataka to distribute food grains at the villagers' doorsteps.'

'That's a tremendous achievement,' I pointed out. 'How did you manage it?'

'It's a question of knowing what one's responsibility is,' Tirlapur said. 'We got the knowledge from the training programme.'

As I said goodbye to each one of them, I looked at them in admiration. Here was a group of ordinary people who had battled the mighty bureaucracy for their rights and finally won. If there is a testimony to what the satellite-based interactive training programme can do in empowering the common man, this is it.

8

IN JHABUA
Watershed

I was going back to Jhabua after some time. I got the feeling that something had changed since my last visit. What had changed, I wondered. Then as I scanned the landscape, I realized what it was. The brown, sandy hills that dot Jhabua's landscape were now green with lush, tall grass that swayed with the wind. Was it because of the watershed project?

I was travelling to Jhabua to get a first-hand experience of how the watershed project had worked. Travelling with me was Dr V. Jayaraman who heads the Earth Observation Systems in ISRO. Dr Jayaraman has a professorial air about him: plenty of erudition, long sentences with difficult syntax, and a tendency to be argumentative. But he knows everything that is to be known about remote sensing; his knowledge is so extensive that it is whispered in ISRO headquarters that remote sensing satellites are often tempted to take commands directly from Dr Jayaraman instead of listening to the on-board computer.

'Tell me something about watershed,' I asked Dr Jayaraman. 'To be honest, I know nothing about it!'

'Let me first tell you what a watershed is,' he explained. 'A watershed is the total land area that drains to a single point. Well, let me put it this way: an area from which rainwater runs off past a single point into a stream, river, lake or an ocean.'

'Are there boundaries to a watershed?' I asked.

He nodded. 'Yes,' he said. 'We call it the drainage divide. What it means is that rains received on one side of the drainage divide do not contribute to the run-off on the other side.'

'What does watershed development do?'

'Well, the idea is to conserve soil and moisture. So that it leads to better agriculture. Let me put it this way: watershed development is about how to utilize land and water resources for maximum production with minimum hazards to natural resources.'

'How do you achieve it?'

'Several things need to be done,' he explained. 'Things like utilizing the land according to its capability, promoting adequate vegetation cover to control soil erosion, harvesting of rainwater in situ, draining out the excess water with safe velocity and harvesting it for future use, and increasing groundwater recharge.'

'A long list, isn't it?' I suggested.

'You think so?' He smiled at me. 'In short, what we have to do is to harvest the available rainfall and recharge the underground aquifers. We have to locate the potential sites and zones for the purpose.'

'How does ISRO do that?'

'Well, sometimes indicators for suitable sites for water harvesting can be identified directly on the satellite data. Let me give you an example. If we construct a check dam at the crossing of a stream and a dyke, the dyke itself will act as a barrier.'

'What does the dyke do?'

'The dyke prevents all water flow—both surface and underground water—on the uphill side. So the result will be a good surface storage reservoir. Now, the question is how to facilitate recharge of groundwater. Let us say, we construct a check dam across a stream following or intersected by a fracture. That will mean that the fracture itself will allow sufficient water to percolate. This will ensure effective recharge to the groundwater aquifers.'

'I still don't understand why this should be done by ISRO's satellites. Can't this be done on the basis of conventional groundwater data?'

'You see,' Dr Jayaraman said, 'satellite remote sensing data provides spatial, multi-spectral and repetitive information. That's why it is such a useful tool for planning for watershed development. We also use conventional data but that is used in conjunction with remote sensing data.'

'What do you use the data for?'

'For several things: delineating ridgelines, characterizing watersheds, assessing the potential, identifying erosion-prone areas, evolving water conservation strategies, and, finally for selecting the sites for check dams and reservoirs.'

The Ambassador car we were travelling in was making weird noises. The driver stopped the car and took a look at the engine. He closed the bonnet and got into the car to restart it.

'Nothing to worry, folks!' he told us in a confident voice.

'There is one more area in which satellite data is useful,' Dr Jayaraman continued. 'It's to monitor the impact of watershed activities.'

'How do you do that?'

'Satellite images depict the status of the watersheds before and after the implementation, thereby indicating the changed status. So we are in a position to know the changes brought about by the watershed programme.'

※※※

We had reached Bori village. It was the headquarters of Action for Social Advancement (ASA), an NGO, which is the programme implementing agency (PIA) in the Watershed Development Project for twenty-five villages covering an area of 10,000 hectares in Udaigarh and Jobat blocks. The idea was that Ashis Mondal, ASA's chief, would take us to Kolyabada, the village where we were supposed to interact with the villagers.

We set out for Kolyabada village in Ashis Mondal's jeep. He was a careful driver, but the road to the village was bad. It rose and fell in steep gradients, much like the landscape through which it curved. We reached the outskirts but even the jeep could no longer negotiate the track: we decided to walk to the village.

It was an hour before sunset. Ashis Mondal showed us the hill where the land had been treated. I could see contour trenches and gully plugs dotting the hillside, and lush grass growing there.

Ashis Mondal explained that the hill, though common land, had been encroached upon by influential villagers and used for unauthorized cultivation. The villagers got together and decided to restore the status of the hill as a common property resource. The encroachers, all influential people of the village and close to the sarpanch, were evicted and the land was brought under treatment. Now the land is protected through social fencing, any violation invites stringent social sanctions from the village community.

The hill slope looked green—a far cry from the bald, brown slopes that greeted one everywhere in Jhabua. As Ashis Mondal explained, the soil erosion had stopped, the run-off of the water had been checked and the topsoil was now retained.

On the way to the village, he showed us the check dam that the villagers had built. It provided irrigation for more than 100 hectares of land. He also took us to see the ponds—some new and others deepened—that the villagers themselves had built.

The watershed activities, he told us, had increased the water table in the village. Before the project, the high seasonal variation in the stream discharges had made it impossible for the villagers to take a rabi crop. But they could take one now, thanks to the water-harvesting bodies. User groups had been formed for each of these bodies.

The sun was setting as we walked down to the village. The smell of maize rotis being cooked wafted up to greet us. It took awhile to get the villagers assembled and we sat there right in the middle of the village road as the fireflies flew restlessly, occasionally lighting up the wizened Bhil faces.

Kaliya, the president of the Watershed Development Committee was there and so was Shri Ram Bhai, a member of the committee. Vesti, Dhuri, Ramli, Dadmi, Thooti and Kundlia—all members of the self-help groups were there, too.

'Now tell us about the watershed works,' I asked.

'Did you see the check dam when you entered the village?' Shri Ram Bhai explained. 'The ponds? Those are the works we've taken up.'

'Do you have a Watershed Development Committee?' Dr Jayaraman asked.

'Yes,' it was Kaliya, the president of the Watershed Development Committee, who spoke now. 'It's the committee of all the self-help groups in the village and of the user groups.'

'What do you do in the committee?' I asked.

'We sit down and plan the works,' Kaliya said. 'All the works you saw were planned by us: that work on the hill, the check dam, the ponds.'

'How did you do it?' I asked.

Kaliya pointed a finger in Ashis Mondal's direction. 'He helped us. He told us what could be done. But he didn't dictate. Not like what the sarpanch does or the BDO. We decided what is to be done and where. The final decision was ours.'

'But you don't know how to build a check dam, do you?' Dr Jayaraman asked.

Kaliya punched the air with his hand. 'Of course, we do. Didn't the Bhils construct all those palaces of the Rajput kings? The ones at Jhabua? Alirajpur? Even the ones in Rajasthan?'

Has a bit of Ashis Mondal's Bengali rebelliousness rubbed off on these Bhils? Or was it the Bhils' pent-up fury of the centuries finally finding expression?

'Those works are different,' I suggested.

Kaliya shrugged. 'Maybe. But we learnt. And of course, Ashis Bhai told us how to do them. We built them. We got the cement, the steel and everything else. Do you know how much we spent on these works?'

'Tell us,' I said.

Pride wavered in Kaliya's voice. 'Something like Rs 6.5 lakh.'

'Why, that's fantastic!' I said.

Kaliya's eyes narrowed. 'Do you know how much the government would have spent for the same works? At least thrice that amount, because the sarpanch would have stolen. The people in the block office would have stolen. And to get water, we would have paid huge bribes.'

'What about maintaining accounts for the works you did?' I asked.

It was Shri Ram Bhai who spoke now. 'We do that, all right. But let me tell you something. Everyone knows how much has been spent. Because we've given our labour. It's our own work, isn't it?'

'Yes, of course,' I said. 'What about the work on the hill slope?'

'That's our work, too,' Shri Ram Bhai said. 'But it belongs to the entire village. We grow grass there for the village. We don't allow anybody to graze cattle there.'

'What happens to the grass?' I asked.

'Part of it we share and part of it we sell,' Kaliya said. 'Do you know that before we started growing grass on that hill slope, we had practically no fodder in the village? We had to go and buy grass from Bori to feed our cattle. Now we sell fodder to other villages.'

'What do you do with the money you get by selling grass?' Dr Jayaraman asked.

'It goes to the Gramkosh,' Shri Ram Bhai said. 'We also give a part of our wages to the Gramkosh.'

'What if somebody says he won't give?' I asked.

Kaliya stuck a finger in the air. 'Oh, everyone gives. Why shouldn't we? The Gramkosh money is our money, isn't it?'

'Do you know something?' Shri Ram Bhai intervened. 'When the watershed works started, they used to deduct the money for the Gramkosh from our wages. Then somebody objected. He said they can't deduct it just like that.'

'What did you do?' I asked.

'We asked Ashis Bhai for his advice,' Shri Ram Bhai continued. 'He said we should discuss about it in the committee. So we sat down and talked about it. Do you know what we did?'

He paused as if to allow us to catch up.

'We decided that the money for the Gramkosh need not be deducted from our wages,' he continued. 'Everyone should be given their full wage and it is for that person to give to the Gramkosh or not.'

'You stopped contributing to the Gramkosh, is that it?' I asked.

'No, no,' Shri Ram Bhai protested. 'We contributed. But only after we got our full wages. Now, we even get receipts for what we give to the Gramkosh. Let me tell you something. Our Gramkosh has got Rs 1.25 lakh. That's a lot of money, isn't it?'

Pride filled his voice like the lush grass on the hill slope outside the village. But then, it was something to be proud of.

'Yes,' I said. 'What do you do with the money in the Gramkosh?'

Shri Ram Bhai's face brightened. 'Oh, so many things! And all good things, mind you. We give loans to the self-help groups, we run a grain bank, we buy seeds and fertilizers, we pay for the teacher in the Falia School, and we're now thinking of building a community hall for the village.'

'That's great,' I said. 'But who decides how the money in the Gramkosh should be spent?'

'Oh, all of us,' Kaliya said. 'The Watershed Development Committee. We sit down and decide.'

I asked them about the activities of the village institutions, and it was Thooti who answered.

'We have two bodies here in this village. One is the Nahi panchayat. That's our self-help group. The one that works. The other is the Moti panchayat. The one that doesn't work.'

'Moti panchayat?' I asked. 'Why "Moti"? Is it because it's loaded with members?'

Kaliya laughed loudly, showing a set of straight, white teeth. 'Oh, that is the regular panchayat. Loaded all right! With the sarpanch and his cronies.'

'Yes, that's what it is,' Thooti continued. 'The one run by the government. We don't know what it does. They don't tell us either.'

'Why? Don't you have a gram sabha?' I asked.

Thooti's face darkened. 'Maybe there's a gram sabha. We don't know about it. We're not invited to the Gram Sabha. It's all for the sarpanch. Vesti, why are you keeping quiet? Aren't you the one who attends the meetings of the Gram Sabha?'

Vesti smiled sheepishly. 'I went only once. That was because the sarpanch's wife dragged me there. What Thooti told you is right. It's the sarpanch who decides whom to call. And he only calls his friends, relatives and family members to the meetings.'

'The family members of the sarpanch?' I asked. 'Are they also members of the panchayat?'

Thooti clucked her tongue. 'Why, yes! Don't you know? There is something now that the panchayat has to have women. So all the women in the sarpanch's family are now in the panchayat.'

'But what about the schemes of the government?' I asked. 'Don't you have these schemes here in this village?'

It was Dhuri who now asked, 'What schemes?'

'Well,' I said defensively. 'Schemes like the IRDP, TRYSEM, Indira Awas Yojana, Employment Assurance Scheme.'

Blank expressions greeted me. Ashis Mondal saw that

too, and started explaining what these schemes meant. I could see comprehension dawning on their faces.

'Oh, those schemes!' Ramli said scornfully. 'They are there all right! The sarpanch and the man from the government decide who'll get what! It's not for us. It's for the friends and lackeys of the sarpanch.'

'Don't tell me that none of you have got money from those schemes?' I asked.

'To tell you the truth, we have,' Ramli said, flushing. 'Do you know what happens? Some lackey of the sarpanch tells us that if we put our thumb impression on a piece of paper, we would get Rs 500. Who doesn't want money? So we signed and got the money. But we found out afterwards that we were supposed to get Rs 2000. And the whole thing was a loan!'

'What happened to the rest of the money?' I asked.

Ramli spat repeatedly on the ground and said scornfully, 'Oh, the sarpanch and the government officers ate it up. We didn't know it at that time. We found out only afterwards.'

'How did you find out?' I asked.

Ramli's expression darkened. 'Only when the fellows from the government came for recovery. They threatened to take away things from our house and our cattle too. We told them that we hadn't got the money but they wouldn't listen.'

'What did you do?' Dr Jayaraman asked.

Ramli groaned. 'What else! We went to the moneylender and borrowed money to pay back the government loan for Rs 2000. After that, we don't go anywhere near a government scheme or government people.'

'But it's your government, isn't it?' Dr Jayaraman pointed out.

It was Thooti who answered, 'We don't know about that! All we know is that the government people come here to take something. If we don't give, they beat us up or send us to jail. We don't want to have anything to do with the government. Certainly not!'

'What about the government works?' I asked.

Thooti made a face. 'Oh, all sorts of government works go on. The sarpanch gets them done. He takes the contract himself for the work in somebody else's name and makes money. We aren't consulted.'

'It can't be all that bad!' I suggested.

It was Shri Ram Bhai, the member of the Watershed Development Committee who now said in a bitter voice, 'Yes, what Thooti said is true. It happens in all the Bhil villages. Do you know what they say about the sarpanch? That he gets up in the morning, climbs his motorbike and goes to the block office to find out what new government work can be taken up.'

'You people were telling me about the other panchayat,' I said. 'Nahi panchayat, isn't it?'

'That's the self-help group,' Ashis Mondal intervened to explain. 'Why don't you tell him about your group, Vesti?'

Vesti brightened. 'Well, that's our own. We're members. We all sit down and decide about things.'

'How often do you meet?' I asked.

'We meet once in a fortnight,' Vesti explained. 'Either during the day or at night.'

'How many nahi panchayats do you have in the village?' Dr Jayaraman asked.

'We have three of them,' Vesti said. 'One is for women and the other two are for men. You see, each family has only one member in the group. We don't allow more than one member from a family.'

'What do you do in the group?' Dr Jayaraman asked.

'We save a fixed amount each month and deposit it,' Vesti continued. 'We use this money for giving loans to members who need it. We also buy fertilizers and seeds with this money.'

'How do you decide who should get a loan?' Dr Jayaraman asked.

Vesti gestured at Dhuri. 'Why don't you tell him, Dhuri? You are the one always talking about loans, isn't it?'

Dhuri spoke up. 'Why are you always after me? All right,

I'll tell. All the members of the group sit down and decide. The first time we met, we decided how much money a member should give every month. When we had enough money, we decided to give loans. The members ask for loans, even for marriages and funerals. We also decide what interest they should pay.'

'What if the person taking the loan doesn't pay back?' I asked.

Dhuri made a face. 'Why shouldn't they? They all pay back. If they don't, we know what to do. We go and sit in front of the person's house and don't leave till the loan is paid back.'

'Don't your husbands object to your contributing to the group every month?' I asked.

It was Vesti who replied with a grin. 'Oh, they don't. If they do, we'll ask them to go to hell. Things have changed. They don't object now. Why should they? It's all for the good, isn't it?'

'Let me ask you a general question, Shri Ram Bhai,' I said. 'Has all this made a difference to your lives?'

He thought for a moment and then said, 'Yes, of course. You see, when we took up that work on the hill slope, we didn't know anything about the soil coming down. Nobody had told us about that. When we got less maize from our lands, we thought it was because we are not putting enough fertilizer. You see, we have a rule here in these Bhil villages—put one bag of fertilizer in one acre of land. So we started putting more and more fertilizer. Do you know what happened?'

I could see that Shri Ram Bhai had a flushed expression on his face. He must be tired, I thought; that was a long speech by the laconic standards of the Bhils.

'Tell me what happened,' I suggested.

Shri Ram Bhai fingered his moustache. 'Our crops got burnt because of too much of fertilizer. That was till Ashis Bhai told us what the real problem was.'

'What was the problem?' I asked him.

'Ashis Bhai told us that the topsoil was getting washed away. We didn't believe him at first. We thought he was talking nonsense. Like people from outside do. What would they know of our land? But then, he did something good.'

'What was it?' Dr Jayaraman asked.

'He selected some boys from our village as volunteers. The boys went to see what was happening in other places. Do you know where our boys went? To Ralegaon Siddhi.'

'Anna Hazare's place?' I asked.

'Yes,' it was Ashis Mondal who replied.

'These boys were lucky,' Shri Ram Bhai continued. 'They went to Dungarpur, too. They even went to a faraway place somewhere in the south. Beyond the Narmada river!'

'To Gulbarga,' Ashis Mondal explained. 'To see the work done by MYRADA in watershed development,'

Shri Ram Bhai smiled. 'Do you know something? The boys came back and told us what they had seen. How the soil can be kept, how the water can be kept, how groups can be formed, and how the groups work.'

'That's wonderful!' I suggested. 'Isn't it?'

Shri Ram Bhai continued as if I had not spoken. 'It opened our eyes. We never knew that we were the culprits. That we were the people who had done bad things to Mother Nature. You see, we Bhils had cut down the trees. We'd cultivated all that forest land. Do you know why we'd done that? Because people told us that if we cut the trees and occupied the forest land, it'll be ours. But now, we know that we'd made a big mistake.'

Shri Ram Bhai rubbed his temple and frowned.

'Listen,' he continued. 'The boys came back from Ralegaon Sidhi and told us about what the people had done in that village. The boys said we must do the same things in our village. But we were hesitant. Do you know why?'

'Tell me,' I suggested.

He waved a hand. 'Because we're poor and uneducated. We've to struggle to make both ends meet. How can we do those big things that the people in Ralegaon Siddhi had done?'

'What did you do?' I asked him.

'We asked Ashis Bhai. He told us that we can do those things here in Kolyabada. Do you know what he did? He made us sit down together and discuss the whole thing over and over again. And all the time, he kept telling us that we had the capacity to do things, that if only we could get together and form groups, we could also do all those things that people in Ralegaon Siddhi have done. So we started forming groups.'

'Now tell me something,' I said. 'Why did you listen to Ashis Bhai? You don't listen to outsiders, not even to the government people. Do you?'

It was Kaliya who spoke now. 'You are right. No, we don't listen to government people. What have they given us? Only a lot of trouble. Ashis Bhai is not like that. He likes us, he wants good things to happen to us. He comes to our village and talks to us, not talk down to us like the government people do.'

'Tell me,' I asked. 'How did you become friends with Ashis Bhai?'

'You see,' Kaliya continued, 'when Ashis Bhai first came here, we didn't trust him. Why should we? We have had such bad experience with outsiders. But he kept coming back. Then he started these health camps for us. He brought a doctor with him. The doctor looked at all of us and gave us medicines.'

'Don't you go to the Badwa any more?' I asked.

Kaliya sniffed. 'No, we don't. Why should we? We now go to Bori. Ashis Bhai has this doctor in his office who looks after us. We even pay for our medicines. You asked me a question, didn't you? I have forgotten it already. How my mind wanders!'

'I asked you why you think Ashis Bhai is a friend,' I said.

'Oh, that!' Kaliya said, grinning. 'It was really the health camps. That made us realize that Ashis Bhai was our friend. We could trust him.'

'You formed these groups and took up all these works,' I said. 'Now let me ask, what have you got from them?'

Kaliya rubbed his chin, as if lost in thought. 'What have we got? A lot, really. You see, the soil does not run down that hill slope any more. Did you see the hill when you came to our village?'

'Yes, he did,' Ashis Mondal said.

'Did you see how green it is now?' Kaliya asked.

I nodded.

Kaliya brightened. 'Do you know that it's the only green hill on this side of Bori? What a lot of grass! But that is not the only thing. With all these dams and tanks we've built, we have much more water now.'

Kaliya took a deep breath and closed his eyes. He had the satisfied look of a man who has achieved a mission.

'Why don't you come and look at our wells?' Kaliya suggested. 'There's so much water now. More water now than we ever had. And what about the water for our fields? We now take a rabi crop because we know that we can get water when we want it. We never took a rabi crop before in Kolyabada village. Oh, no!'

'Let me ask you something else,' I said. 'Do you still go to the moneylender?'

Kaliya groaned. 'No, no. Not any more. Why should we? We have our nahi panchayats. And the Gramkosh. And what we have there is our own money. We can take from it, but we have to give it back. Otherwise, there won't be any money there the next time we want it, isn't it?

'Yes, of course,' I agreed. 'Do your members pay back their loans?'

Kaliya nodded. 'They do. Maybe some Bhil has a problem paying back. The group listens to him and finds out whether his problem is genuine. If it is, he gets more time to pay back his loan. But let me tell you, everyone pays back in the end.'

'What about migration?' I asked.

It was Shri Ram Bhai who answered. 'Some people still go. But it's much less now. I ask you: Why should we go to faroff places to work when we have work waiting for us right here on our doorstep? And it's our own work, isn't it?'

'Yes,' I agreed. 'Let me ask you something else. What about the sarpanch? Does he tell you what to do?'

'Not in these matters,' Shri Ram Bhai continued. 'Certainly not. But he comes to us for votes, though. We now ask him what he is going to do for us if we give him our vote. It was not like that before. Those days, he took away our votes. We never asked him any questions.'

'Do you invite the sarpanch for your meetings?' I asked. 'The meetings of Nahi panchayat?'

Kaliya said defiantly, 'Why should we? If he is a member, he'll come. And he'll sit down with us like any other member. In our meetings, he's the same as any one of us. Not more, not less.'

I started talking to the women about their self-help group.

It was Vesti who explained. 'Our group is very active. It has twenty-one members. We regularly give our savings to the group. And the group gives loans.'

'Don't your men shout at you for giving money to the group every month?' I asked.

Vesti made a face. 'Why should they? It's not their money. It's our earnings, isn't it?'

'Yes, of course,' I agreed.

She smiled at me. 'But I know why you ask. Those things happened before we formed our group. That time, our men used to take away our earnings and drink. It doesn't happen any more.'

'You mean to say that the men don't drink any more?' I asked.

Vesti gestured towards Kundlia and said, 'Let her tell you. She was the leader for that. Why don't you tell him about it, Kundlia?'

Kundlia's lively eyes flitted from one thing to another and finally focussed on me. 'They do drink. But only on special occasions. Not like before. And because they don't drink regularly, there's none of those fights and looting.'

'Do the men allow you to take decisions in your group?' I asked. 'Or do they decide for you?'

It was Vesti who answered. 'Oh, no! We decide things in our group. Even in the committee, we say what we feel like.'

'Do your children go to school?' I asked Vesti.

'Yes. All of them. We have even started a school. A Falia school. We pay the teacher out of the money in the Gramkosh.'

'Let me ask the women something,' I said. 'Do you think all these things—the Nahi panchayat, the committee, growing grass on the hill slope, the dam, the tanks—have they made your life better than what it was?'

There was a silence. Were the women thinking it over?

It was Vesti who finally said, 'Yes, of course. We have work now. We earn more. We don't have to go to faraway places in search of work. The work is right here in our village. And we sit down, talk about things and take decisions. We don't go to the moneylender. The Nahi panchayat gives us loans.'

'Has all this given you a sense of power over your life?' I asked.

Vesti frowned. She obviously didn't understand my question. One of the volunteers told her in the Bhil language what I meant.

Vesti nodded vigorously and her ornaments made a clanging noise. 'It has. We now know that we can do things for ourselves.'

'Tell me something, Kaliya,' I asked. 'Ashis Bhai is not going to be around all the time. The project is going to end soon. Do you think you can manage on your own?'

Kaliya looked down and produced a double chin as he did so.

'I don't know,' he replied finally. 'Wait, I think we can. You see, Ashis Bhai is telling us all the time that he is going away one day, that we have to do things for ourselves. Yes, I think we can manage.'

'We have been preparing the villagers for that,' Ashis Mondal pointed out. 'Of course, they'll manage. I have no doubt about that.'

The fireflies were tired and so were we. It was time for

us to leave. We said goodbye to the villagers and wished them luck.

Ashis Mondal took us to see a biogas plant that had been put up in one of the houses. The biogas plant was nice, but what was nicer was the smile on the faces of the Bhil women. Their cheeks bulged with well-being; was it because they did not have to blow at the fire to keep it going?

As we walked back to Ashis Mondal's jeep, I could make out the hill blurred against the darkness. But I could see in my mind's eye the hill slope covered with lush green grass that had brought so much cheer to the Bhil lives. And I could hear the water in the village stream gurgling away happily: it sounded like the ringing laughter of the Bhil women—the women who had taken charge of their lives.

※※※

The National Centre for Human Settlements & Environment (NCHSE) is the programme implementing agency for the watershed project for five villages in Rama block and fourteen villages in Jhabua block. It was in Makankvi, one of the villages with NCHSE that we had fixed up our interactions with the villagers.

Vikas Kulashreshtha, the person in charge of NCHSE in Jhabua district, had fixed up our interactions. Makankvi was only 6 kilometres from Jhabua town on the road to Para, but we had to fork off the main road. The gravel road we had to use now petered out into a stony nothingness very soon. We left our vehicle and walked about a kilometre to Makankvi village.

We were expected. The villagers were waiting for us near the newly constructed check dam on the outskirts of the village. A mat had been spread on the bund of the check dam. I was introduced to Gajri, Jelu and Hakri who had been waiting for us. We started talking to them.

'Tell me, Gajri,' I asked. 'Has this watershed project been useful to you people?'

Gajri nodded. 'Very useful,' she said. 'You see, more than anything else, we've learnt such a lot from it.'

'What have you learnt?' I asked her.

'Many things! Like how we had been greedy and destroyed everything by cutting down trees and clearing the forest.'

'Why, what happened?' Dr Jayaraman asked.

'Well, the soil on the hilltop got washed away. And that's why there was no water in our wells. And the water under the ground was all gone.'

'When did you discover this?' Dr Jayaraman asked.

'Well, only at the time we prepared the action plans. Do you know what we used to think before? When the rains didn't come and our crops went bad year after year, we thought it was the rain-god who was punishing us!'

'Why should the rain-god punish you?' I asked.

'For something bad our Bhil ancestors had done to the rain-god long, long back. They had insulted him.'

'Is that so?' I commented.

Gajri said with an embarrassed smile, 'No, no! Now we know it was not because our ancestors had insulted the rain-god. It was because the Bhils had cut down trees and cleared the forest that there was no water.'

She sighed a deep sigh. A sigh that seemed to travel down her entire body, down to her toe-rings.

'Do you know what had happened to our wells?' Gajri continued. 'They were going deeper and deeper every year. But we didn't know why. NCHSE people told us why it was so.'

'Why don't you tell him, Gajri,' Jelu suggested, 'about what they had told us about grazing our cattle in the common land?'

'Yes, yes,' Gajri said as if galvanized. 'They also told us that we shouldn't be grazing our cattle so much on the common land because it is bad for the soil.'

'What did you do?' Dr Jayaraman asked.

'Well,' Jelu said, 'they told us how we could set things right. We listened to them and we liked what they told us. Do you know something? NCHSE took us to places where we could see how the watershed works were done.'

'Where did you go?' I asked Jelu.

She sat up. 'We went to Kalyanpura, here in Jhabua block. We also went to Udaipur in Rajasthan.'

'What did you see?' I asked.

'Oh!' Jelu's voice was euphoric, 'Wonderful things! We saw how ordinary people like us have built watershed works and what a lot of good it has done for them. After we came back, we agreed to do it here. We sat down and talked to all the villagers; they agreed.'

'What about the Watershed Development Committee?' Dr Jayaraman asked.

Gajri intervened to say, 'We are all in the committee. All our group members. So we started doing all these watershed works. Like the one here; we built this dam.'

'But how is it different from the works that the panchayat does?' Dr Jayaraman asked.

It was Hakri who said in an outraged voice, 'Oh, god! How can you say that? The sarpanch does these works without asking anybody in the village. And all bad work! They are no good. Not even one paisa of benefit to anybody.'

'Why should that be?' I asked her.

She pursed her lips. 'Because the sarpanch does these works to make money for himself. Why should he care if the works are no good?'

'Doesn't he ask you people?' I asked.

'Oh, no!' Hakri said indignantly. 'We're not consulted. The works are planned between the sarpanch and the government officials. You should see the roadwork that he has done. So terrible! And all kinds of false bills! The sarpanch and the government people looted all that money!'

'How is your watershed works different?' Dr Jayaraman asked.

Hakri's voice rose. 'Because these are our own works. We all decide what the work should be and where it should be done. We do the work ourselves.'

'So who makes the money?' I asked.

'Oh, nobody!' Gajri exclaimed. 'I can tell you that nobody is allowed to make any money! The volunteer does the accounts. The accounts are explained to everyone and if anybody has any doubt, they shout. Do you know how much these watershed works would have cost if they had been done by the sarpanch?'

'Tell me,' I suggested.

Gajri clucked her tongue. 'At least, three times of what they have cost now! And they would have been in the wrong place, I can tell you that.'

'Have things improved for the village after these watershed works?' Dr Jayaraman asked.

'Very much so,' Gajri continued. 'There is so much water now. You should see the water in our wells. It has gone up. You can almost put your hand down and touch the water. It's so high!'

'What about the crop?' Dr Jayaraman asked.

Gajri smiled. 'It's much better now. Do you know that we even take a rabi crop? And also vegetable crops. But the best thing is that there is so much more work in our villages now.'

'Has migration stopped?' Dr Jayaraman asked.

'Not totally,' Gajri replied. 'Some people still go. But it is much less. You can count their number on fingertips. Now tell me, why should we go to faraway places when there is so much work here?'

'That's true,' I agreed. 'Do the Bhils have enough to eat throughout the year?'

Gajri sighed. 'Yes, now we have. But we had problems in the past. What we got from our fields didn't last us for the whole year. We had to go to the moneylender and borrow grains. And what a lot of interest the moneylender charged! And we had to sell our next crop to him very cheap. No, that kind of thing doesn't happen anymore.'

'How many self-help groups are there in this village?' I asked.

'Three of them,' Gajri said. 'There are about ten to thirteen members in each group. The groups are functioning for the last three years.'

'Tell us about your group,' I suggested. 'What does the group do?'

'Well, each member saves Re 1 per day and a maximum up to Rs 50 in a month. We deposit the money in the group's accounts. We maintain the bank passbooks.'

'What happens if a member doesn't save money?' I asked.

She raised her eyebrows. 'Why shouldn't she? We all do. All of us work and get wages from watershed works, don't we?'

'I am sure you do. Do you also get money from other sources for your group?' I asked.

'Yes, yes. We get money from the bank, four times of what we save. We also get money from the collector. I know the scheme, but I can't pronounce it.'

'Under the DWACRA scheme,' Kulashreshtha explained.

Why did the Bhil women have a problem with pronouncing the name of the scheme? Couldn't the government have given a simpler name to a scheme meant for illiterate village women?

Gajri nodded. 'Yes, that's the scheme. We give loans to our members, for seeds, for fertilizers. We also give for schemes that can make money. Like making of strings or flower garlands.'

'Do the members repay their loans?' Dr Jayaraman asked.

'Yes, they do. If a member has some problem, she comes to the group and tells us why she can't pay back. We put our heads together and find a solution. Maybe, we give her a few more days to pay back.'

'How often does the group meet?' Dr Jayaraman asked.

'Once in a month. More often, if there's something important to be done.'

'Do all the members attend the meetings of the group?' Dr Jayaraman asked.

'Why not?' Gajri answered my question with a question. 'Of course, they do. It's also their group, isn't it?'

'What do you do if a member doesn't attend the group meetings?' Dr Jayaraman asked.

Gajri grinned. 'Well, we know what to do. We go to the house of the absent member and demand refreshment or breakfast or tea. That's her punishment for not attending the meetings of the group.'

'Do you think the group has done good things for the Bhil women?' I asked.

She nodded. 'Oh yes. Now we save money. We never did that before. Do you know what used to happen when we needed money? We used to go to the moneylender and borrow money after putting our thumb impression.'

There was a faraway expression in Gajri's eyes. Was she thinking of those times, the harrowing times?

'It looked so easy at that time,' she continued after a while. 'But there was no end to it. To repay the moneylender, we had to sell him our crop at a very cheap price. But that was not all. We had to migrate so that we could earn cash and pay back the moneylender's dues. At that time, we never realized that we were doing something wrong.'

'When did you realize that?' Dr Jayaraman asked.

Gajri clucked her tongue. 'You won't believe it! Only when the people from NCHSE came to our village and talked to us. They told us that if we saved money, we didn't have to go to the moneylender. They asked us to organize groups. Then we could borrow from the group itself.'

'Do you go to the moneylender now?' I asked.

She shook her head. 'No, not now! Why should we? We have our group. If the group doesn't have money, we go to the bank, borrow money and give loans to the members. Or the group takes a loan from the Gramkosh and pays it back.'

'How much money have you saved so far?' I asked.

It was Hakri who replied, 'About Rs 1.5 lakh for all the three groups in the village. Well, Gajri forgot to tell you something. The reason why the moneylender cheated us was because we didn't know how to read. He'll write something and say that's what we owed him. We had to accept whatever he told us.'

Gajri nodded. 'That's true. We were cheated because we're illiterate. We Bhils are very scared of the written word. That's why we avoid places where reading or writing is needed.'

'Are you still afraid of the written word?' Dr Jayaraman asked.

Gajri held up her hand and her bangles fell on top of each other with a clink. 'No, not any more. Not after the groups have started working. We are still illiterate, but we are not afraid now. We ask somebody to explain it to us. We now understand many more things that we didn't understand before.'

'Do you wish that you were educated, Gajri?' Dr Jayaraman asked.

'Yes,' she said wistfully. 'Education is very important. You see, all this work has made us realize how important education is. We want our children to go to school. Particularly, our girls.'

'Why do you say that?' I asked.

Gajri groaned. 'Because there is something wrong with our girls! They don't want to continue in school. They want to get married. We have this Bhagoria Hat. The girls run away and get married. There's nothing we can do, can we?'

'Can't you tell your girls how important education is?' I suggested.

She made a face. 'We do. We tell them that this is the time for going to school and getting educated. But they won't listen.'

'Well, there's something else,' Jelu intervened to say. 'The Bhils say that it is useless to educate a girl because she would marry and go away to another house.'

Hakri nodded. 'That's true. Let me tell you something. There are problems when a Bhil girl gets educated. An educated girl refuses to get married to Bhil boys who don't have education. Then the girls have to be married off forcibly. In that case, the bride price is much higher.'

Gajri clucked her tongue. 'Oh, don't listen to Hakri! That happens only in a few cases! The main thing is the taboo. That a Bhil girl shouldn't be allowed to go to school after she comes of age. Do you know why? Because we don't have a school in each Falia. So a girl has to walk very far to go to a school. And the parents don't want that once the girl comes of age.'

'What's to be done, then?' I asked her.

She chewed on her lower lip. 'We have to open more and more schools. In each Falia, there should be a school. So that girls don't have to walk long distances to go to a school. Do you know something? Our group has already started a Falia school.'

'That's very good!' I pointed out.

'Oh, we now understand the importance of education,' Gajri said. 'In our village, we see to it that every child goes to school.'

'Tell me something, Gajri,' I asked. 'This project—watershed, self-help groups, and watershed committees—has it given you a lot of power?'

She gave me a thoughtful look and then, smiled. 'Well, let me see. Yes, the project has done things for us. We don't starve in the lean days. We don't have to migrate to faraway places in search of work. There is work here, our crops are better. We don't have to go to the moneylender; we save money now. We can take a loan from the group and start something. You see, I've taken a loan for a smokeless chulla and a small grocery shop.'

'What about control over your life?' I asked.

She pursed her lips. 'I don't know! But I can tell you one thing. For us Bhil women things are better than before. Our menfolk listen to us now. They never used to do that before. They consult us. Do you know what my husband said? He said that I must contest in the panchayat elections. He would not have suggested it before.'

'Are you contesting, Gajri?' I asked.

She stuck a finger in the air. 'Why not? But the elections are postponed now. Because there is some order, I believe.'

Kulashreshtha explained to me that the court had stayed the panchayat elections on some technical ground.

'Do you know something?' Gajri said. 'The project has opened our eyes to many things we didn't understand before. Like the importance of education, the importance of watershed, and the importance of savings. There is something else, though.'

'What?' I asked.

She smiled. 'It has taught us that we shouldn't do certain things. Like drinking liquor. You see, the Bhils drink a lot and they do a lot of bad things when they are drunk. Like fighting with each other.' Gajri's voice was disapproving. 'Committing murders. Attacking vehicles on the road and looting them.'

'What did you do?' I asked.

'We all sat down and decided that drinking was bad. That nobody should drink.'

'Did that work?' Dr Jayaraman asked.

'The villagers have kept their promise. They don't drink anymore.'

'That's very creditable,' I said.

'Let me tell you something else. We also decided that we'll not give bribes to anybody, not even to the police. And it has worked. We don't give bribes now.'

'Then how do you get your work done with the government?' Dr Jayaraman asked.

'We get them done without paying bribes,' Gajri said. 'Some things get done, some don't. But we've decided that we aren't going to pay bribes. Now, you've asked so many questions and we've answered them. Will you come to my house and see my smokeless chulla?'

We went to Gajri's house. It was very dark inside. It took me awhile to get used to that darkness, and then, I saw the smokeless chulla. Gajri proudly showed us her grocery shop.

Nothing great really—some bidis, some cigarettes, some packets of salt, sugar, and matchboxes. But in the Bhil setting, a great entrepreneurial effort indeed.

When Gajri bade us goodbye, she made us promise that we should come again to see how they were doing. As I made my way back to the world outside through agricultural fields that now smiled at me with a rabi crop, I told myself that I had to come again. If only to see the progress of these Bhil women in their emancipatory endeavour.

※※※

We tried to get in touch with Deepak Joshi a number of times. He was the project director of the Centre for Bio-Conservation and Development (CBCD) which is the project implementing agency (PIA) for the watershed project in seventeen villages of Thandla block. But he never seemed to be around in Thandla. His office told me that he was away in some place where we could not reach him. Why was Deepak Joshi so mysteriously peripatetic?

We managed to fix up a visit to Badi Dhami village without the intervention of Deepak Joshi. CBCD was the PIA for the watershed project in Badi Dhami village as well as the adjoining village. Badi Dhami was about 8 kilometres from Thandla, the tahsil headquarters, and 45 kilometres from Jhabua.

As we approached the village, a church wove into view. There were a number of pucca buildings in the vicinity of the church.

'Are there a number of Christians in the village?' I asked the driver.

'Yes,' he said. 'The majority of the people in the village are Christians.'

'What are those buildings next to the church?' Dr Jayaraman asked.

'Oh, that's a school. The church people run that school.'

We went to the office of the village panchayat. It seemed

I was in luck. It was the scheduled day of the fortnightly meeting of the Watershed Development Committee, and the members were there in full force. And by some strange coincidence, Deepak Joshi was there, too.

The room was swarming with Bhil women; there were a few men, too. Deepak Joshi introduced me to Paskeli, Bhuri Behn, Anna, Tara, Kali and Saroj, Shri Patik and Laloo.

'Aren't you all members of the Watershed Development Committee?' I asked.

It was Anna who answered my question. 'Yes. That's why we are here today. It's our meeting day.'

'Has your committee taken up watershed works here in the village?' Dr Jayaraman asked.

'Yes, it has,' Anna said. 'Many things. A dam just outside the village, a big plantation of bamboo and also cultivation of grass in the common land.'

'That's a whole lot of work,' I pointed out. 'But tell me what you did to make the watershed programme work.'

Tara gestured to Bhuri Behn. 'Let Bhuri Behn tell you that. After all, she is the one who went to Ralegaon Sidhi. Didn't you, Bhuri Behn?'

Bhuri Behn said with an embarrassed smile, 'I don't know why everyone picks on me because I went to Ralegaon Sidhi! All of them know what we have to do. All right, all right. Let me explain.'

But I could see that she was quite eager to explain.

'Well, we have to level our land,' she said in a high-pitched voice. 'Cultivate only across the slopes. We have to do bunding. What is that bunding called, Deepak Bhai?'

'Contour bunding,' Deepak Joshi explained.

Bhuri Behn looked gratefully at Deepak Joshi. 'Yes, that's what it is. We have to plant trees on those bunds. We have to collect the monsoon rain and store it. We have to control the grazing of our cattle and do it on rotation basis. And we have to grow grass or fodder. Isn't that it?'

Bhuri Behn, in her own simple way, had described the scope of watershed project more succinctly than I would ever do.

'Tell me something, Bhuri Behn,' I asked. 'Why was your land so bad that it needs to be treated?'

She pursed her lips. 'Oh, we didn't know that time. Now we know. People from CBCD told us. We didn't believe it when they told us that. But it's true!'

'Why, what did they tell you?' Dr Jayaraman asked.

'That, about fifty years back, our district was covered with thick forest. At that time, whatever the Bhils needed, was given by the forest—our fuel, our fodder, timber for our houses, everything.'

'But I don't see any forest now,' I said. 'What did you do to the forest?'

'Terrible things,' it was Paskeli who explained. 'Do you know what happened? Some political leaders told the Bhils that they should cut down the trees and cultivate the land, so that the land will be theirs, like it was in the past. So the Bhils cut down the trees and started cultivating the land. What a stupid thing to do!'

'Do you realize that now?' I asked.

Paskeli frowned. 'Yes, we do. But only now. See what happened! The rains became less and less. And all the time we thought that it was because our ancestors had gone and fought with the God of clouds. That's why the God was punishing the Bhils by sending less rain year after year.'

'Do you know now what really happened? Dr Jayaraman asked.

'Yes. All this happened because the Bhils cut down the trees. That's why there are no tree roots to hold water. The water under the ground also went down. You should have seen our wells that time! The water was so deep down!'

'Is that so?' Dr Jayaraman asked.

Paskeli nodded. 'It was terrible. When it rained, all the water just ran off because there was nothing to hold that water. You should have seen our hills. They were so bald!'

'Some parts of Jhabua district are still like that,' I suggested. 'Isn't it?'

'Yes. Because the Bhils made all kinds of mistakes. There

was very little land for the cattle to graze. We let the cattle graze wherever there was vacant land.'

'What do you do now?' I asked.

'Well, we don't let our cattle graze in the common land. We now use the common land for something else.'

'For what?' I asked.

'In the common land, we grow bamboos,' Paskeli continued. 'We have also planted grass there. To grow fodder for our cattle.'

'Do you have enough fodder now?' Dr Jayaraman asked.

It was Kali who replied. Was she the fodder expert? 'More than we need. We even sell fodder.'

'How do you manage that?' I asked her.

'All the people in the village cut the grass together and collect it in a place in the village. The Watershed Development Committee calculates the requirement of fodder for all the families in the village and the grass is distributed according to that.'

'But you said you also sold a part of it,' I said, 'isn't it?'

'Yes, that's true,' Kali agreed. 'We have decided that half of the grass that we get from the common land should be sold and the money deposited in the Gramkosh. We do that.'

'That's fantastic,' I said. 'Now, has your committee taken up watershed works in the village?'

'Quite a few,' Kali said.

'Have you people benefited from these watershed works?' I asked.

'A lot,' Kali assured me. 'Let me tell you what has happened. The water level in the village has increased; you should see our wells, they are full now, not like before. And you should see the common land that we have treated. Old roots of the trees that the Bhils had cut fifty years back are coming back to life.'

'That's very impressive,' I suggested. 'Now, what about your crop? Is there an improvement?'

'Yes. We even take a rabi crop now. And we have managed to get our own lands treated also.'

'Is a lot of work going on in the village now?' I asked.

'Yes.' It was a male voice and I looked around for the owner of the voice. It was Shri Patik, member of the Watershed Development Committee. 'There is plenty of work here in the village.'

'Do the villagers still go out on migration to Rajasthan and Gujarat?' I asked.

'Migration is much less now,' Patik said. 'Very few Bhils go now. Tell me, why should they migrate when there is work here in the village?'

'What about those bad months of May to July?' Dr Jayaraman asked. 'What is the position now that you people are implementing the watershed project?'

Patik fingered his moustache. 'You're right,' he told Dr Jayaraman. 'Those were the bad months for the Bhils because, at that time, the kharif crop is still not harvested. So we had to go to the moneylender to borrow grains. Do you know what the interest used to be?'

Dr Jayaraman shook his head. 'No,' he said.

'It was 100–150 per cent,' Patik pointed out. 'That, too, for a season! That kind of thing doesn't happen anymore after we've started implementing the Watershed Project.'

'Why?' Dr Jayaraman asked.

'Because we've opened a grain bank right here in the village,' Patik explained.

'How did you find the money for it?' Dr Jayaraman asked.

'Let Laloo tell you that,' Petik suggested. 'He is the one who really started the grain bank.'

'From the funds of the Watershed Project meant for the entry point works of the CBCD,' Laloo pointed out. 'Deepak Bhai gave it to us.'

'That's interesting,' I intervened to say. 'Now, tell me, what did you people have to do to start the grain bank?'

'First, we sat down and made the rules,' Laloo explained.

'Who are "we"?' I asked Laloo.

'The members of the Watershed Development Committee

and the self-help groups of Badi Dhami village,' Laloo said. 'Who else?'

The fact that Laloo was a member of the Watershed Development Committee had slipped my mind.

'What kind of rules?' Dr Jayaraman asked.

'Rules about the villagers taking out grain loans,' Laloo explained. 'We made a rule. That each family can take up to 50 kilos of grains in two instalments. When the loan is returned, it should be of the same quality as what was borrowed. We have even given passbooks to the villagers. So that entries can be made about the loans taken and the repayment made.'

'Sounds like a proper bank,' Dr Jayaraman observed. 'Isn't it? Do you charge interest?'

Laloo nodded. 'Yes, we do,' he said. 'But the rate is much less than what the moneylender would've charged. Let me tell you how it works. If the grain is returned within three months, the rate of interest is five per cent. It is ten per cent if the grain is returned between three to six months. If it's after six months, the rate of interest is fifteen per cent and also a penalty. What the penalty should be is decided by the Watershed Development Committee.'

'You haven't answered Dr Jayaraman's question,' I pointed out. 'Do the Bhils still go to the moneylender for borrowing grains after you've started implementing the Watershed Project?'

Laloo protested, 'No, no! Why should they? They come to the grain bank. And now that the project has increased the availability of water and the crop yield, the Bhils are in a position to return the grain loan easily and in time.'

'Well, that reminds me,' I said. 'Do you people still go to the moneylender to take money loans?'

It was Paskeli who answered my question. 'No, we don't,' she declared. 'We are all members of some self-help group or the other. The groups lend money for all kinds of things; even for medical things. We also give loans to Bhils who are not members of the self-help groups. But mind you, at a higher rate of interest.'

'What do these moneylenders do now?' Dr Jayaraman asked.

'We've taken away their business,' Anna said with a laugh.

The women suggested that we should have a look at the check dam that they had recently built. It was just outside the village, they said.

We went off to see the check dam. It was outside the village, all right; but a good 1 kilometre from the panchayat office. It had rained when we had been talking and the path to the check dam was now slushy. But the Bhil women walked ahead in long strides. Gaiety marked their steps and their ornaments tinkled merrily.

We finally reached the check dam, but not before Deepak Joshi had slipped and almost fallen into the nullah. Walking down slushy village pathways was obviously not one of his strengths.

Pride wavered in the voices of those Bhil women as they vied with each other to tell me how much the dam had cost, how many days it had taken to build and how many acres it was going to irrigate. The construction was of good quality and the location of the dam very strategic. I told them that and they sounded pleased.

'Do you know that we went to Thandla to buy the cement and steel for the dam?' Paskeli said in a voice tinged with pride. 'And we had to bargain and bargain till we got the things very cheap.'

'Suppose,' I said, 'this work had been done by the sarpanch. How much would it have cost?'

'A lot more,' Anna suggested. 'At least, twice or thrice of what it has cost us. That's because the sarpanch and the people from the block office would have eaten up most of the money.'

'Would they have consulted you?' I asked.

'No, never!' Anna said. 'There are so many works that they did. They never asked anybody, they didn't even ask our menfolk. Nobody consulted us.'

'Why not?' I asked.

'Because we are women,' Paskeli said. 'Before this project came, nobody bothered to ask us anything. Not even our menfolk.'

'What happens now?' I asked.

'Well, they have to,' Paskeli's voice was triumphant. 'Isn't it? Because we are members of the Watershed Development Committee.'

Paskeli started laughing. Peals on peals of it. Other Bhil women joined in the laughter; all of them—Bhuri Behn, Anna, Tara, Kali and Saroj.

It was time for us to leave. We said goodbyes, and as we rounded the corner near the church building, the laughter of the Bhil women was still ringing in my ears. Was the laughter one of liberation?

9

THE HIMALAYAS IN GARHWAL
Pilgrims' Progress

There was nothing in common between Protima Bedi and Ambi. Protima was a celebrity every eventful day of her tragically brief life. She shot into fame when, barely out of her teens, she streaked across a crowded beach in Juhu. She went on to become an accomplished Odissi dancer. She was beautiful with large, liquid eyes; courted controversy whenever she could; and was unfailingly flamboyant.

There was nothing remotely flamboyant about Ambi. He looked after ISRO's accounts and did it rather well. But much like the accounts he kept, he was faceless, and if he could manage it, nameless, too. Stories are still told in ISRO headquarters of how, every time Ambi was complimented for a thing well done, he would do his best to disappear into the woodworks. Ambi was a very reticent man.

Death brought them together. They were part of the team that was going on a pilgrimage to Kailash–Mansarovar. One fateful evening, the team camped at Malpa in Uttaranchal Himalayas, and by next morning, it had been buried alive in a landslide that took place during the night. In the landslide at Malpa in 1998, 300 people died, including sixty pilgrims. It was the worst ever landslide in recent times.

Such landslides are fairly common in the Himalayan terrain. They kill hundreds of people every year, damage properties and block communication links. They occur due to the complex geological setting with contemporary adjustments to varying slopes and relief, heavy snowfall, rainfall and anthropogenic effects. The landslides are of two kinds, deep and shallow. Structural features such as joints, faults, and bedding planes induce deep landslides while shallow landslides are more influenced by the nature of the weathered mantle.

ISRO has prepared landslide maps of the Himalayas. While mapping the area, the accent was on covering pilgrim and tourist routes. The routes that were covered consisted of Rishikesh–Rudraprayag–Chamoli–Badrinath road, Rudraprayag–Okhimath–Kedarnath road, Chamoli–Okhimath road, Rishikesh–Uttarkashi–Gangotri–Gaumukh road, Pithoragarh–Khela–Malpa road, Shimla–Rampur–Sarahan–Sumdo road, Shimla–Bilaspur–Kulu–Manali road and Dalhousie–Chamba–Brahmaur road.

I travelled on the Rishikesh–Rudraprayag–Chamoli–Badrinath road to see for myself how the mapping was done. Travelling with me was Dr V.S. Hegde, the head of Disaster Management System in ISRO. Dr Hegde looks like an absent-minded scientist who has lost most of his hair in trying to solve intractable problems in experiments lasting through the night in stuffy, overheated laboratories. Absent-minded he may look, but he has a first-rate mind coupled with the physical strength to trudge in inhospitable terrains for days on end to map difficult areas.

We left Rishikesh early in the morning. People were up and about, but a thin mist hung over the town. The river Ganga was a muddy brown, and as we climbed up, it veered off from the road to our right. We soon crossed Chandrabhaga River which joins Ganga later. I asked Dr Hegde, 'Tell me how does ISRO come into disaster management?'

'ISRO plays a very important role,' Dr Hegde explained. 'You see, the combination of remote sensing, communication and meteorological satellites helps in managing disasters. It is the synergy between the remote sensing and communication satellite capabilities in managing disaster situations.'

'What about landslides?' I asked. 'How do space systems help?'

'In several ways. They help in respect of preparedness. Space systems provide mapping of areas that are prone to landslides. They help in preparing guidelines for land use and engineering structures. They also help in laying down a risk assessment framework.'

'What about prediction?'

'Yes, space systems help in that, too. You see, space systems provide modelling of landslide processes. They also provide the slope and soil stability information.'

'What about relief and rehabilitation?'

'ISRO's inputs play a very crucial role in relief operations. They provide details of areas affected by landslides. They also give an estimate of relief road and communications. In respect of rehabilitation, they give details of resilient structures and terrain farming.'

'Tell me, what causes these landslides in the Himalayas?'

'There are two causative factors. The first has to do with preparatory factors such as geology, topography and environment. The second are the triggering factors. Such as rainfall, earthquakes, volcanic eruptions, and land use changes.'

'What about the topography of the Himalayas?' I asked. 'Is it prone to landslides?'

'You see,' Dr Hegde explained, 'in the Himalayas, it is the lithology that controls the landslides. We see that the softer rocks such as phyllites and shale are more prone to landslides when compared to harder rocks such as quartzite and granite.'

'What about the geological structures?' I asked. 'Are they important?'

Dr Hegde nodded. 'Yes. Landslides occur commonly along the major thrusts and faults in the Himalayas. That's why faults in general are considered significant geological structures for landslide hazard zonation.'

The road forked off to the right at Kailash Ashram. Ganga now flowed right next to the road.

'One more thing,' Dr Hegde continued. 'The attitude of different beds with relation to topographic slope has an important bearing on landslides. When the bedding dips or the master joint plane coincides with topographic slope and if it is towards the road, it triggers landslides.'

'What about geomorphology?'

'That's important, too,' Dr Hegde explained. 'It is because some of the important geomorphic elements give us a clue to the future landslides in that area. You see, topographic parameters like slope aspect and slope morphology play a significant role in landslide.'

'How's that?'

'Because, a slope greater than 25 degrees is significant for landslides. So what we do is to identify categories of slopes beyond 25 degrees from the toposheets. All these are mapped from the satellite data by drawing profile along critical cross-sections in the toposheets.'

'Are landslides influenced by present-day land use?'

'Yes, that also has an important bearing on landslide hazard zonation and mitigation measures. You see, the satellite data has the capability to directly record features such as different density of vegetation, the rocky exposures and agricultural lands.'

'What about the topsoil?'

'That's important, too. Particularly in case of shallow landslides in the Himalayas. Weathering is important in triggering shallow landslides. So what we do is to identify different degrees of weathering in the field in a grid-based mode. Then we extrapolate it to the inaccessible areas using geomorphology, drainage, lineament density, depth of soil, and land use and land cover as controls. We map the drainage network of the area from the toposheet and update it using the satellite data.'

'What really triggers the landslides in the Himalayas?' I asked.

'Rainfall is the major triggering factor. In the Himalayas, landslides occur after an intense spell of rain. So what we've done is to collect the historical rainfall data for the last several years from the nearest rain gauge stations. We've plotted the data in a graph to understand the rainfall behaviour in a particular year.'

'What kind of methodology do you adopt for zonating landslide hazards?'

'We use a model addressing expert knowledge combined with a statistical technique. How each parameter of the terrain and categories within the parameters respond to landslide susceptibility are considered and ranked as per expert opinion. The ranks are then translated into weightages using software. And then we classify the hazards. We've indicated six hazard classes in the maps. They are: severe, very high, high, moderate, low and very low.'

We had just crossed Saur village on the road when Dr Hegde made the driver stop the car. We both got out.

Dr Hegde explained, 'I want to show you the site of an old landslide.'

He pointed to a site right on the road. I could see that a massive landslide had taken place at that site. It had left behind heaps and heaps of debris.

'When we were mapping this area and preparing the hazard zonation maps,' Dr Hegde explained, 'we found that this was a site that might be reactivated due to triggering factors. So we recommended safeguards to be taken.'

'How have you classified this area in your hazard zonation map?' I asked.

'It's classified as very high in the hazard class,' he said.

Dr Hegde pointed to a place a little beyond the old landslide.

'That's the area,' he explained, 'we've classified as only high. That's because chances of landslide occurring there is less than in the place we're standing on right now.'

We walked back to the car and resumed our journey. We passed through Devaprayag town.

'This is where the rivers Bhagirathi and Alakananda meet to create the Ganga.' Dr Hegde said. 'From here, all the way down to the plains, it is only the river Ganga.'

'How do you characterize this area in your hazard zonation map?' I asked.

'High,' Dr Hegde said.

The road turned right and the river Alakananda stayed with us on our left. At Gobindkoti village, Dr Hegde took me off to show the site of an active landslide. It looked as if the entire hill slope had come down in a solid mass. There was debris strewn all over the place.

'Let me ask you something,' I said. 'How do you validate your hazard zonation maps?'

'You see, the entire database is created in a digital form. If the cursor in the computer screen is placed at any given point on the landslide hazard zonation map, resource information becomes available for checking and validation. The maps we prepare are validated by comparison with ground conditions.'

We passed through a fairly large-sized settlement.

'This is Srinagar town, home to Garhwal University,' Dr Hegde explained. 'Once upon a time, it was the capital of Garhwal state. But around 1803, the raja of Garhwal abandoned it because of damages caused by recurring floods.'

I looked at Dr Hegde in bafflement. How can floods affect towns in a hilly terrain like this?

'How is that possible?' I asked. 'Aren't the towns here much above the rivers flowing in the valleys?'

'What you say is true in the conventional sense,' Dr Hegde explained. 'But here, we're talking about landslide-induced floods. You see, when the debris from the landslide rolls down, they make temporary dams in the rivers. That creates temporary lakes. But soon, the backwater pressure of the lakes exceeds the retention capacity of the barrier. Then the accumulated water gushes downstream with mighty force. That inundates towns and human settlements which are otherwise quite safe.'

'Floods in this hilly terrain!' I said. 'I wouldn't have believed it!'

'Let me tell you that this is particularly true of the area we are presently passing through. The area along the river Alakananda. There have been many such floods in this area.'

We had reached a big town. I could see a burbling river meandering through the hills and joining Alakananda near the town.

'This is Rudraprayag,' Dr Hegde said. 'This is where the river Mandakini meets Alakananda.' It was fascinating—the vague green of Mandakini merging with the sludge-hued Alakananda.

'Where does Mandakini originate?' I asked.

'Near Kedarnath. At a height of about 12,000 feet.'

We moved on. I asked Dr Hegde, 'We were talking about the landslide-induced floods in Alakananda. Tell me, do these floods cause a lot of devastation?'

He nodded. 'Yes, they do. Let me tell you about the Alakananda floods of 1970. You see, there was a massive cloudburst in July 1970 in Kunwai Khall ridge. It so happened that the watersheds of the streams emerging from Kunwai Khall area had been made barren in the preceding years.'

'Why, what was the reason?'

'Large-scale deforestation. There was incessant rain that year and it caused many landslides. The debris from the landslides blocked Alakananda by forming a high dam of 20–40 meters. As a result, water ceased to flow in the river. A police constable on duty saw this and alerted the people. Because of the constable's warning, about 400 pilgrims who were travelling to Badrinath saved themselves, by running up the slope.'

'That was lucky!'

'Others were not as lucky. The torrent of floodwater just engulfed a convoy of about thirty-one vehicles and they were swept away without any trace. There was a lot of devastation caused by the flood.'

We were passing through a beautiful landscape. The

gentle slopes of the hills all around us were clothed in foliage of yellow and rust, interspersed with an occasional deep red. This was the autumnal Himalayas at its finest display of colour.

'The rains seem to be the triggering factor for the landslides,' I said. 'Are there other reasons?'

Dr Hegde nodded. 'Yes. But they are mainly anthropogenic. Things like deforestation, overgrazing, forest fires, and construction of roads.'

'A long list of culprits, isn't it?' I suggested.

'Yes. But the main culprit is deforestation. The problem of deforestation can be traced back to the British days. When the British came to this area, they took the forest away from local control and declared it as reserve areas. This was as early as 1911.'

We passed through a fairly large-sized town.

'This is Karnaprayag,' Dr Hegde informed me. 'This is where the river Pindar meets Alakananda.'

'Where does Pindar originate?' I asked.

'In the Pindar glacier high up in the hills. That's about 11,000 feet. But there is something interesting about the river Pindar. It travels a very long way down the hills before it meets Alakananda. About 115 kilometres.'

'It's a long river,' I agreed. 'Well, we were talking about the British coming into this area. What did they do to the forest?'

'Something terrible! You see, the wealth of the rich forests tempted the British to exploit it for revenue. That's how the commercial exploitation of the forests in this area started, and it went on for more than seventy years till it was stopped by the Chipko Movement.'

'Tell me, what was the damage like?'

'Very extensive. A lot of damage was done to the terrain. Let me tell you what used to happen at that time. Because there were no roads in the interior, the timber that was felled was transported in the rivers. That's why only those trees which could float in the water were cut. But then, the roads were built. And the commercial exploitation of timber increased. Even trees like oaks which had not been touched by the lumber's axe, were also cut.'

The landscape around was changing. The gorges were becoming deeper as were the ravines and crevices. There were patches of terraced paddy fields so stunningly green that they took one's breath away. We were approaching a big town. I could see another river meeting Alakananda on the periphery of the town.

'This is Nandaprayag,' Dr Hegde declared. 'It is so named because this is where the river Mandakini meets Alakananda.'

Soon, Nandaprayag was behind us and we resumed our conversation.

Dr Hegde continued. 'Prof. Purohit, a brilliant botanist who teaches in Garhwal University, says the roads in this region were not built out of necessity, but for the commercial rate of return.'

'There's something in what Prof. Purohit says.'

'You see, construction of these roads was partly responsible for the landslides in the Himalayas. It destabilized the hillside and produced huge amounts of debris. The debris rolled down the slopes into the valleys. In many areas, roads were dug without providing for proper buttress wall. After the roads were built, landslides became common during the monsoon.'

'What happens when there is deforestation?'

'In the Himalayan setting, forests are important because they influence the environment and moderate the climate. They also maintain the soil cover, regulate water and purify the air. Tree species like deodar, chir, pine and oak were cut for fodder, fuel, timber, furniture and making of agricultural implements. So when the protective forest cover was lost, it led to soil erosion and landslides. The result was increasing amounts of silt and floodwater down the streams.'

'You talked about overgrazing, didn't you?'

'Yes,' Dr Hegde said. 'You see, the cattle population in

these parts far exceeds the carrying capacity of the forests and pastures. Overgrazing has almost the same impact as deforestation.'

'In what way?'

'Overgrazing of pastures results in a deterioration of the vegetation cover. Cattle consume palatable species. They are replaced by unpalatable shrubs, herbs and grasses. On the whole, the overall impact on the vegetation cover is bad.'

The landscape had changed once again. The hills, which were such a comforting presence so far, had become mountains. Even the colour of the Alakananda had changed to a subtle aquamarine as we approached a big town.

'This is Chamoli,' Dr Hegde declared. 'This is where the road branches off on one side to Okhimath and finally to Kedarnath, the pilgrimage centre.'

'I remember reading about a big earthquake in Chamoli,' I said. 'Wasn't there an earthquake here?'

'Yes, there was a big one, in March 1999. It caused large-scale damage to human lives and property here.'

We passed Chamoli town and the road climbed up.

'This area is called the Pipalkoti area,' Dr Hegde said. 'Let me show you something interesting here.'

We got down from the car at Kauriya village. Dr Hegde showed me a site of an old landslide. It was right on the road.

'What a big landslide!' I exclaimed. 'I wonder how much devastation it caused!'

'There was a lot of devastation. But what's remarkable is that it has five human settlements sitting on top of this site. Kauriya, Batula, Sirkot, Mayasasur and Digoli. But the worrisome thing is that this site can be reactivated anytime due to triggering factors.'

'How does this area rate in your hazard zonation mapping?'

'Parts of it are high while other parts are moderate. But what is of concern is that two of the villages by the roadside—Kauriya and Batula—are in the area that we rate high.'

The landscape had changed once again. Pines had started

appearing in increasing numbers, dotting the landscape. There were fields now planted with a red flower.

'What is that red flower, Dr Hegde?' I asked.

'Oh, that is madua,' Dr Hegde explained. 'That's a millet-like crop favoured by the locals here. We are now in the Valley of the Gods. This range is a part of Nanda Devi Biosphere Reserve.'

We passed through another town.

'What is this town?' I asked.

Dr Hegde said, 'This is Joshimath. This is historically very important because this is where Adi Shankaracharya attained his enlightenment. And this is the place which houses the winter hibernation of Lord Badrinath.'

The view outside the car window was breathtaking. There were rows upon rows of majestic hills, with the Alakananda flowing through them like a silver thread. Soon we passed through Vishnuprayag where Dhauti Ganga meets Alakananda, and Hanumanchatti, where Kairon Ganga originating in Neelkanth Peaks joins the Alakananda. Further on, I saw a large number of trekkers going off the road.

'Where are these trekkers headed?' I asked.

Dr Hegde peered out of the car. 'Oh, they are going off to the Valley of Flowers,' he said.

We were now approaching Badrinath, our destination. Alakananda had become a bluish green. There were green meadows everywhere. The valley nestling Badrinath was beautiful. Several peaks surrounded the valley, standing tall and inflexible, like sentinels of spirituality. The mountains were in a semicircular formation, with the Nar and Narayan peaks, and the pride of place taken by the majestic, snow-capped Neelkanth shimmering in the dewy sunshine.

I closed my eyes. The picture that came unbidden to my mind was the colourful facade of the Badri temple, the tinkling bells and the frenzied chants of the faithful resting their travel-weary limbs in the verandas of the precincts. I could see Lord Badrinarayana in black stone, representing Lord Vishnu. The same Lord Vishnu who had done his penance right here in Badrinath, and seeing him do his penance out in the open, Goddess Mahalakshmi had assumed the form of a Badri tree to provide him shelter from the harsh weather.

Our car came to a stop because an avalanche had cut-off the narrow strip of pathway to the temple. I could see an old woman, her body bent double over a staff, walking briskly on the gorge path to the temple with a determination that only a staunch devotee can muster.

I nudged Dr Hegde and said, 'Do you see that old lady with the bent back? It's for people like her that ISRO would have made the journey safe with its landslide maps.'

There was a sudden gust of wind that blew away my words; maybe all the way to the Neelkanth peak.

'Eh?' Dr Hegde said, with his mouth open, looking more like the absent-minded scientist that I always thought he was.

10

THE SUNDARBANS
Protecting the Ecosystem

When Goddess Ganga descended from the heavens, she would have split the Earth with her mighty torrent. Lord Shiva, asked to save the Earth, tamed Ganga's torrent by trapping her in his ash-smeared locks. So the river Ganga, as it appeared on Earth, was a heavenly braid; an immense rope of water unfurling through the parched plains of northern India. But before it met the sea, the heavenly braid came unstuck: it separated into hundreds, maybe thousands, of tangled strands.

That is how Amitav Ghosh describes the Sundarbans, the estuary of the river Ganga. In the Sundarbans delta, the river's channels are spread across the land like tangled strands of Shiva's heavenly braid, creating a terrain where the boundaries between land and water are always changing. Some of these channels are wide waterways, so wide that one shore is a distant blur from the other; others are thin, only a few 100 meters across.

The Sundarbans comprises an archipelago of islands that stretches for almost 300 kilometres. Some of these islands are vast, while some are no larger than sandbars. Some of the islands have survived through recorded history while others have just been washed into existence. In any case, the geological formation of the Sundarbans is of comparatively recent origin. Till a few 1000 years back, the whole tract was under the sea. The deposits of debris and the formation of the Sundarbans delta came about when the river Ganga changed its course.

The history of human settlement in the Sundarbans is also relatively recent. It dates back to the treaty of 1757 signed by Mir Jafar, through which the lands in the 24 Parganas district were ceded to the East India Company and eventually became the *jagir* of Lord Clive. That was precisely the time when the harvesting of timber from the forests in the Sundarbans began. Thereafter, there has been a continuous history of forests being reclaimed for agriculture. In the last century alone, more than 750 square kilometres of slightly more than 4000 square kilometres of the forest area has been diverted for agricultural purposes. Small wonder, then, that the ecosystem of the Sundarbans is seriously threatened.

※※※

Dr Jeyaram and I travelled to the Sundarbans to get a first-hand impression of things there. Dr Jeyaram, a brilliant application scientist, heads the Regional Remote Sensing Service Centre (RRSSC), Kharagpur; an ISRO unit which has done impressive work in the area of protecting the ecosystem of the Sundarbans.

We travelled to the Sundarbans the old-fashioned way, like people from the outlying areas do when they travel to Kolkata to work in the city. We took the commuter train from Dhakuria, a south Kolkata station to Canning. Canning is the most important town in the Sundarbans, but it wears a forlorn look. But it was not always like that. Lord Canning, the then viceroy of India, planned it to be a substitute for Calcutta (modern Kolkata). In Canning's time, they wanted a new port and eventually, a new capital for Bengal; the river Hooghly was silting up and it was believed that its docks would soon be choked with mud. So Lord Canning sent out teams of planners and surveyors to scout for a new place. And on the banks of the river Matla, they found the place, named it Canning after the viceroy, founded a town and built a port. But they had reckoned without the river Matla. One day, in the year 1867, precisely fifteen years after Canning came into being, Matla hurled itself on the town and in a matter of hours the town was all but gone; only the bleached skeleton remained.

Over time, Canning has built on that bleached skeleton, but the forlorn look persists. It is now a jumble of narrow lanes, cramped shops and mildewed houses. The bazaars end in a causeway that leads away from the town towards the river Matla. We hired a boat at Canning. Like everything else in Canning, it was decrepit. It was a diesel steamer that had seen better days, now adapted for the tourist trade, with rows of plastic chairs lined up behind the wheelhouse. Dr Jeyaram and I made the brave walk up the steep gangplank and settled down on the plastic chairs behind the noisy wheelhouse as comfortably as we could and waited for the boat to start.

The boat finally started with its ancient engine spluttering and hammering. I started talking to Dr Jeyaram.

'I believe ISRO has done a lot of work on the protection of the ecosystem in the Sundarbans,' I asked. 'What exactly have you done?'

Dr Jeyaram explained, 'ISRO has been mapping and monitoring the wetlands of the Sundarbans delta by using satellite imagery. On the basis of the maps and analysing the ground truth, ISRO has prepared a plan for the conservation and protection of the ecosystem of the Sundarbans.'

'Why, what's wrong with the ecosystem of the Sundarbans?' I asked.

'You see, the ecosystem in the Sundarbans is an unusual one. The region is one of high tidal amplitude. And fluctuations too. Cyclones and storms are fairly common. So are tidal waves.'

'Do the tides travel inland?'

'They do. The tides reach as far as 300 kilometres inland. The currents of these tides are so powerful that they reshape the islands almost daily. Some days, the water tears away entire peninsulas. At other times, it throws up new shelves and sandbars where there were none before.'

Dr Jeyaram has a way with words, I thought.

'The Sundarbans delta is a dynamic ecosystem,' Dr Jeyaram continued. 'It is continuously being created by the process of accretion of new land through tidal sedimentation, as well as erosion of banks. This leads to the formation of new islands. During the peak high tide, the saline seawater inundates the islands. There is no flow of fresh water in the river systems. So the islands in the Sundarbans are subject to changing salinity, soil texture and tidal action as well as biotic pressures.'

'What about the forest in the Sundarbans delta?' I asked.

'The forest in the Sundarbans delta is unique. For one, it is the largest mangrove forest in the world. It accounts for 60 per cent of India's total mangrove cover. You see, when the tides in the Sundarbans create new land, overnight mangroves are born. They spread so fast that they cover a new island within a few short years. Have you seen a mangrove tree?'

'Only from a distance.'

'Mangrove leaves are short and leathery. The branches are gnarled. The foliage is very dense. Visibility is short. The air in the mangrove forest is still and fetid. There is nothing pretty about the mangrove forests, but they provide a variety of ecosystem services.'

'Like what?'

'These ecosystems store large amounts of inorganic and organic nutrients, which are washed into the mangroves. They also process huge amounts of organic matter, dissolved nutrients, pesticides and other pollutants that are dumped into mangrove areas. So they act as biological wastewater treatment plants by removing toxic elements.'

'What about fisheries?'

'I was coming to that. The mangrove forests in the Sundarbans act as a natural nursery for a variety of commercially important prawns, crabs and finfish. That is because they provide abundant food and shelter for them. They also provide food, roosting and nesting sites to a large variety of birds. They also support many tropical aquatic and terrestrial organisms by enriching the fertility of estuarine water for production of planktons.'

'What exactly is happening to the mangrove forests?'

'Awful things. Because of continuous biotic pressure, the mangroves are experiencing habitat loss. There are changes in species composition. There are shifts in dominance. There is loss in biodiversity and threat to survival. Activities like bunding, erosion and deposition have caused changes in tides and currents which, in turn, have affected the mangroves.'

'These mangrove forests have tigers,' I said. 'Don't they?'

'Yes, that's true. The mangrove forest in the Sundarbans is also the only one in the world with a tiger population. It is home to the Royal Bengal tiger. But unfortunately, people have cleared the forest for cultivation. They also poach the spotted deer in the forest. This has created serious problems.'

'Problems?' I asked. 'Like what?'

'You see, the spotted deer is the favourite food of the tigers here. Since people have been poaching the spotted deer relentlessly, there is no food for the tigers. They are hungry.'

'That's funny,' I pointed out. 'Amitav Ghosh talks of hungry tides in the Sundarbans. You talk of the hungry tigers!'

He protested. 'But that's how it is. Without food to eat, the tigers here in the Sundarbans will grow old at a tender age. That's why they are thinking of an old age home for the tigers.'

I laughed out aloud. 'Why, that's even funnier! An old age home for tigers!'

Dr Jeyaram protested. 'But that's true. The Forest Department is seriously thinking of it. They have even set apart 45 hectares of land in Jharkhali Island.'

'How strange!'

'But we should realize that this has serious implications. Once the tigers don't find food in their natural habitat, they'll start foraying into neighbouring villages. They'll target cattle or domestic animals. Even human beings.'

'That's a serious matter,' I said. 'Tell me about the people who live in the Sundarbans.'

'Well, the population here is very heterogeneous,' Dr Jeyaram explained. 'They are basically migrants. They came to settle here in the period after independence.'

'Where did they come from?' I asked.

'Mainly the displaced people from the East. In the Sundarbans, we have a very high representation of the minorities and other disadvantaged social groups.'

'Tell me, what do people in the Sundarbans do for a livelihood?'

'Well, they are either farmers or fishermen, or both. The farmers cultivate paddy, betel, chilly and vegetables. The fishermen venture into the sea to catch fish. Fishery is a major means of income. The fishermen use plank-built canoes or mechanized boats and nylon gears of fishing. Major types of fish caught include catfish, hilsa, prawns, rays and crabs.'

'What about agriculture?' I asked. 'Is that the main occupation?'

'Yes, a large part of the population depends on agriculture. But agriculture here is mono-cropping: only one crop a year. And that too, totally rain-fed.'

'What do they grow?'

'They grow paddy, but a particular kind of paddy called boro rice.'

'Do all of them own land?'

'No. Almost half of the households are agricultural labourers. The terrain here being what it is, there's not enough land to go around.'

'So what do they do for a livelihood, if they can't find employment in agriculture?'

'They go for fishing. For those who are economically backward, fishing consists of collecting Bagda prawn seeds.'

'How does that work out?'

'Rather badly, I am afraid. You see, this business of collecting Bagda prawn seeds is overexploited. It is risky, too. It means standing in waist-deep or deeper water for hours on end. It also involves many health hazards. Such as skin diseases and bites from various water species.'

'Why, that's terrible!' I exclaimed.

'Wait!' Dr Jeyaram said. 'I haven't told you the whole story yet. There are other risks as well. It often means being eaten up by crocodiles and sharks, while standing there in waist-deep water. That's why the male members of the families refuse to do this work as they are perceived as the primary earners of the household.'

'Then, who does this kind of work?'

'Mainly women and children. In fact, many children are kept out of school to do seed collection along with their mothers.'

'Tell me something. Why should people migrate to the Sundarbans to be exposed to such harsh livelihood patterns?'

Dr Jeyaram shrugged. 'I suppose they would've been worse off elsewhere.'

※※※

Our boat slowed down and I could see that we were approaching a town. It was crowded and I could see several hotels on the seafront. The beach was beautiful and remarkably clean. And totally free of hawkers.

'That's Bakkhali, our first destination,' Dr Jeyaram said, pointing towards the town. 'The sunset here is spectacular. And in the mornings, you can see the sun rise behind the windmills. It's really beautiful.'

I could see Sundari trees dotting the creek. The creek was a hub of fishing activity.

'Do you see that narrow path disappearing into the casuarina forest bordering the beach?' Dr Jeyaram continued. 'That goes to the temple of the Bana Bibi. She is the goddess of the local people of the Sundarbans.'

Our boat landed at Bakkhali fishing harbour with a thud. A lot of activity was taking place in the harbour. I could see landings of prawn, hilsa, pomfret and catfish. The person we were supposed to meet in Bakkhali harbour was Bapi Gayen. We looked for him everywhere, but could not find him. We finally found him driving a tough bargain for several basketfuls of catfish. He finally bought what he wanted and came down to talk to us.

Bapi Gayen was a young man of about thirty. He sported a striped blue and white T-shirt that was very trendy. He had a long, aquiline nose, even teeth and prominent ears that stood out like cauliflowers.

'Did you strike a good bargain?' I asked.

'What bargain?' Bapi Gayen said with a sneer. 'How can one strike a bargain in this wretched place? Everyone is so desperately poor in these parts! If you are out to make a little money, the other fellow loses. Perhaps he'll go hungry if you make money at his expense.'

'What do you do for a living?' Dr Jeyaram asked.

'I vend fish,' he said. 'Didn't you see what I was buying just now? Eh, did you think I was buying all the loads of catfish to eat at home? I am not that rich!'

'Do you make enough money from fish vending?' I asked.

He shook his head. 'Not really! But that's because everything here in the Sundarbans is going from bad to worse. Let me tell you that the fish landings have come down drastically.'

'Why is that?' I asked.

There was a deep frown on Bapi Gayen's face. 'Because fishing has become a risky business. Particularly in the high seas. And let me tell you that's where the good catches are.'

'Why has it become risky?' I asked. 'You mean to say it was not risky before?'

'Why is it risky now? That's because of what happens to you when you are out there in the sea. There are all kinds of things now! Typhoons, cyclones, hurricanes, storms! And they come without any warning to anyone. So there is no fishing on those days. You see, accidents can happen. One can even lose one's life!'

'But these things were there before, weren't they?' I protested.

He said ruefully, 'True, they were there at that time, too. But they happened once in a blue moon. One could still do decent fishing on most days of the year. Fishing was less risky then. Let me tell you that there's something seriously wrong with what's happening now.'

'Why do you think things have changed?' Dr Jeyaram asked.

'Well, I'll say the sea is angry with us. It wants to destroy us. Have you seen how the sea is intruding into the land over the years?'

'Yes,' Dr Jeyaram agreed.

Bapi Gayen's voice rose. 'I used to be in Canning when I was younger. I've seen the river Matla in those days. It was like a huge waterway. It was full of water. And that too, sweet water.'

'And now?' I asked.

Bapi Gayen frowned. 'It is like a thread! And no sweet water now; only salt water!'

'Well, things are changing,' Dr Jeyaram said. 'Let me tell you why these changes are taking place. Do you go to the mangrove forest?'

'Yes, I do.' Bapi Gayen said. 'You see, one can't make a living from fish vending alone. There are those lean months when there are no fish landings. That time, I go to the mangrove forests and collect foliage.'

'What do you do with the foliage?' I asked.

'Well, here we use the foliage as fodder for cattle and goats. So I sell them.'

'Have you noticed that the mangrove forest is going down?' Dr Jeyaram asked.

'Of course, I've noticed that. How can the forest remain the same when everyone here sponges on it?'

'What do they do?' I asked.

'Well, nowadays, they use it for making charcoal. Also for tannin. For paper also, I believe. The other day, I saw somebody collecting huge quantities of mangrove for making dyes and chemicals. How can the poor mangrove forest survive all that extraction?'

'But you also use the mangrove forest, don't you?' Dr Jeyaram asked.

His voice was rebellious. 'Yes, I do. But I collect only green leaves and fruits for fodder. I don't destroy anything, do I? You know what the people in the Sundarbans do? They use the wood for fuel, furniture, and even construction. Let me ask you again: Can the poor forest survive such onslaught?'

'It certainly can't,' Dr Jeyaram said. 'But do you know what good things will follow when people don't destroy the mangrove forest?'

'All I know is that in that case, it'll give me more green leaves and fruits for fodder.'

'Well, if you conserve the mangrove forest, it'll act as a barrier against typhoons, storms, cyclones and hurricanes.'

'Will it do that?' he asked with hope in his voice. 'In that case, it'll make fishing a lot less risky, and my business will improve.'

'Not only that,' Dr Jeyaram said. 'It'll also stop the sea from intruding into land. The thing that you fear so much.'

'Will it really do that?' I asked. 'To tell you the truth, that is my worst fear. Sometimes, I have nightmares about it. Tell me something. Do you really think these small, gnarled mangroves are capable of doing all those things you say they can?'

'Yes,' Dr Jeyaram said with authority in his voice. 'They certainly can do these and lots more.'

Bapi Gayen looked relieved. 'Can I go now?' he asked. 'The fellow I bought this catfish from has more of the same stuff. Now that some time has elapsed and there are no customers in sight, he may sell it to me much cheaper.'

He ran off in search of that bargain.

✳✳✳

We went to Bakkhali town and met Nagen Biswas. He had migrated to India from erstwhile East Pakistan a long time back. He had a nervous, angular face with balding hair and a gamchha—a kind of hand towel—that was tied to his forehead. He was dressed in a frayed dhoti and an old-fashioned banian that hung down to his elbows.

'I hear that you came down here from Pakistan,' I asked. 'When was that?'

He furrowed his brow. 'But that was a long time back. I don't remember which year it was. But I do know it was before Bangladesh came into being.'

'How old are you?' I asked.

He grinned sheepishly. 'All of sixty-five years.'

'How were things when you first came here?' I asked.

'Things were much better. There were not that many people in the islands. There was much more forest that time. One could make a decent livelihood. The fish catches were good, there was sweet water in the rivers, and, the sea didn't advance into the islands so much.'

'You mean to say that there were no high tides?' I asked.

'Of course, there were high tides. But the sea was far away. And the tides didn't enter our homes like they do today. Nowadays, the sea advances right into the islands in great fury. And this is happening very regularly, I can tell you that.'

'Do you know why?' I asked.

He knitted his brow again. 'I am an ordinary man. How will I know that? I am only telling you what happens in the islands nowadays. There's a great deal of erosion taking place. Also, at the same time, there is a lot of accretion in several areas. Very strange, isn't it?'

'Are you worried about that?' Dr Jeyaram asked.

He nodded. 'To tell you the truth, I'm worried. I honestly think, one day, this bloody sea is going to take away all of us. It's no longer safe here in the islands. Things are changing very fast.'

'How are things changing?' I asked.

'I know at least one thing. When the high tide is at its peak, the saline seawater comes into the islands and floods everything in sight. That's why there are these embankments here. Have you seen them?'

'Yes, I've seen these embankments,' I said. 'The islands look like forts.'

'Let me tell you what happens during the ebb tide,' Nagen Biswas continued. 'At that time, the seawater recedes from the interior of the islands. But when the water recedes, it takes away a lot of soil along with it.'

'You mean the topsoil?' Dr Jeyaram asked.

'I don't know those fancy words. But I suppose that's what the ebb tide takes away. Let me tell you that is how these new channels are formed. After some time, these channels keep advancing into the islands. And that is how you have these muddy banks.'

'But what are you worried about?' I asked.

'About the sea coming in so aggressively inland: it's very dangerous. I'm also worried about erosion taking place everywhere.'

'But don't you have the mangrove forests as a protection against that?' Dr Jeyaram asked.

He smiled ruefully. 'Well, that is no longer the case. Most of them have been destroyed. I should know that, shouldn't I? Once upon a time, I used to go to the mangrove forests quite regularly to collect honey and firewood.' He shook his head now. 'But no longer!'

'What's the reason?' I asked.

'Because, nowadays, to get honey and firewood from the mangrove forest is next to impossible. You see, most of the forest has been destroyed either for habitation or for other reasons. People have been greedy!'

'Do you know that because the mangrove forests have been destroyed' Dr Jeyaram suggested, 'the sea is advancing inwards?'

He scratched his head. 'Is that so? Well, I never thought of that! Is that why the sea is advancing into the islands? But let me tell you something. One day, the sea is going to take all of us away.'

We took leave of Nagen Biswas.

※※※

The person we met next in Bakkhali town was Ravindra Jana. He looked about thirty, with a neatly trimmed moustache and an intense expression. He wore a slightly frayed brown bush shirt with white checks.

'What do you do for a living?' I asked.

'I'm an agriculturist,' he said. 'But only a marginal farmer. I own a very small piece of agricultural land.'

'What's the kind of crop you grow?' I asked.

'I grow boro rice. But I can take only one crop a year. That's because all our lands are rain-fed.'

'Is that so?' I asked. 'But tell me, do you get enough rice to feed you for the entire year?'

He shook his head. 'No chance! In fact, I get very little rice. Certainly not enough to last for the entire year.'

'So what do you do?' I asked. 'Do you do other things as well?

He nodded. 'Yes, I do other things. Mainly, I collect some timber, firewood and honey from the mangrove forest.'

'Does that give you enough money?' I asked.

'Not much, but it is very risky. There are tigers there in the mangrove forest. So you see, one can't go very deep into the forest. Then, there are snakes and crocodiles.'

'Do people get killed in the mangrove forests?' I asked.

'Yes, they do get killed by snakes, tigers and crocodiles.'

'Why is that?' I asked. 'Can't you keep an eye on them?'

Ravindra Jana shook his head. 'No, that's difficult. You see, the mangrove foliage is very thick. You can't see very far there; visibility is very poor. One doesn't know what is lying in wait for you at the next step!'

'Do you know that there is a lot of destruction of the mangrove forest?' Dr Jeyaram asked.

He nodded. 'Yes, I do know that. Even the little piece of land that I own was converted by clearing the mangrove forest. My father did that.'

'But do you know that clearing the mangrove forest is bad for the island?' Dr Jeyaram said solemnly.

'Yes, yes. I heard that on the radio the other day. We're now trying to see that people don't destroy the mangrove forest. I hope we'll succeed.'

Dr Jeyaram nudged me to say that we were getting late for our next appointments. We were making the trip to Frazerganj by boat. We took leave of Ravindra Jana and got on to our boat.

✳✳✳

Exactly 100 years ago, Andrew Frazer, the lieutenant governor of Bengal fell in love with a beach at one end of the Sundarbans. He was so enchanted by that beautiful beach that he built a bungalow nearby, surrounded by swaying coconut palms. In course of time, the bungalow was swallowed by the advancing sea; today, nothing remains of Frazerganj's colonial past. But the beautiful beach that Andrew Frazer loved so much still remains, as enchanting and inviting as it had been 100 years before.

Frazerganj is now a bustling fishing town with a harbour. I could see a lot of activity taking place at the harbour itself. Boats were unloading loads of catfish and there was a great deal of animated bargaining going on.

It was in the harbour that we met Somen Mitra. He looked about forty, with a crop of light hair that swayed in the sea breeze, a moustache and deep-set eyes with a piercing gaze. He sported a collar-less T-Shirt that was blue once upon a time, but had now been bleached into a nondescript off-white.

'What do you do for a living?' I asked.

'I own a fish boat,' he said laconically.

I smiled at him. 'That's nice,' I said. 'You should be making a lot of money, isn't it?'

He frowned. 'Not really. You see, I employ a large crew. About forty of them. So out of the total money I make from the boat, I keep 40 per cent. The rest is shared amongst the crew.'

'How much money do you think a person here makes from fishing?' I asked.

He scratched his head. 'Difficult to say. But let me see. During the peak season, a person here makes about Rs 40,000. That, in any case, is his earning for the entire year.'

'That's not much!' I said. 'Why is that?'

He made a face. 'That's because the fish landings are getting to be less and less with every passing year. We don't go very far into the sea. There are risks if you do so. You get caught in the storms, cyclones or strong winds. Then, of course, there'll be casualties.'

'So where do you fish?' I asked.

'As close to the coast as possible. You see, everything here seems to be changing. The main thing is that the fish catch is dwindling. At the rate the fish catch is going down, I don't know what will happen to people like us.'

'Do you know what has happened to the mangrove forests?' Dr Jeyaram asked.

'What has happened to them?' Somen Mitra answered the question with a question. 'Well, I won't know! I never go that side, if I can help it. They say there are man-eating tigers there. But there is talk that the mangrove forests have been cut down in a big way.'

Dr Jeyaram explained patiently. 'These mangroves support fisheries in a big way. That is because the mangrove forest is a natural nursery for fish. So if you protect the mangroves, the fish landings will certainly be much better.'

'Is that so?' Somen Mitra said cheerfully. 'In that case, I'll make more money. For God's sake, I never realized that these mangroves, with those tough, leathery leaves, were so useful!'

✳✳✳

We met Sheikh Jamaludeen in his house in Frazerganj town. He looked like a man who made his living with his hands. His unruly crop of salt and pepper hair sat uneasily on his head. He had a moustache that needed trimming and stubble that was several days old.

'What do you do for a living?' I asked.

He frowned. 'Well, I have no land. So I depend on subsidiary occupations. I do firewood collection in the mangrove forest here. But it is not an easy job, I can tell you that.'

'Is it because of the tigers?'

He nodded. 'Partly yes. But believe me, I have not set eyes on a tiger so far. Though I have heard them growl.'

'Is that so?' I asked. 'But that must be terrifying!'

His face clouded. 'Yes, it was. But nowadays, I go to the mangrove forests only to a limited extent. And that too, to those islands where there are no tigers.'

'That's safe, isn't it?' I pointed out.

He laughed. 'Yes. But the problem is that in the islands where there are no tigers, the mangrove forests are being cleared. That's not a good thing for people like us who depend on the mangrove forests for a living.'

I asked. 'Has the sea affected the forest here?'

He pondered over my question for some time, and then said, 'Yes, the sea has intruded into the forest in some places. Yes, it has.'

'Has it decreased the density of the forest?' Dr Jeyaram asked.

'No, not at all. The density of the forest has not decreased because of the intrusion of the sea. It has decreased because of human beings. They have cleared it. That is why the sea is entering the islands, causing such destruction.'

'What is being done about it?' Dr Jeyaram inquired.

'Well, I know that your organization has pointed out how clearing the mangrove forest is bad. Now people are talking about it. I think as a result of that, there is some improvement. Most people have stopped clearing the forest.'

Dr Jeyaram told me that we better leave. We took leave of Sheikh Jamaludeen.

※※※

As our boat chugged back to Canning, I knew that what needs to be done for the ecosystem is to spread environmental awareness. ISRO had already taken the crucial first step in the right direction. The wetlands in the Sundarbans had to be preserved. That alone can help save the delicate and threatened ecosystem.

11

IN THE GANGETIC PLAINS
Reclaiming Sodic Land

If you happen to be travelling through the Gangetic plains of Uttar Pradesh on a summer day, heat comes at you in waves that shimmer in mid-air. If you scan the landscape, you see the waves rising, phoenix-like, from barren lands that look a rabid yellow, as if bathed in artificial light. It looks as if these lands have not been cultivated for years together, and in places where they have been, the crop is so pathetic that you wonder why anybody took the trouble of cultivating them at all.

These are the sodic lands of Uttar Pradesh, which form 7 per cent of the state's net cultivable land. They are either barren or marginally productive. They have high concentrations of exchangeable sodium in which finer soil particles are dispersed, and water and air cannot penetrate them, creating highly alkaline conditions. Sodic soils are toxic to plants, and injurious to human and animal health.

The pity is that much of these sodic lands are owned by poor people. Large areas of such sodic lands were distributed to the landless people by the government. They could not be brought under cultivation because huge resources were required for their reclamation, which the poor who were given the land could hardly afford. The result is that these lands are either barren or barely productive.

The Uttar Pradesh Sodic Lands Reclamation Project (UPSLRP) seems to have changed all that. Now that the project is almost complete, the results are impressive. The project has turned barren sodic soil into fertile, arable land, and the beneficiaries have participated in the programme's implementation from start to finish. Over its seven-year implementation, this poverty-based project has reclaimed about 1,90,000 hectares in twenty-one of the twenty-eight districts of the state most affected by sodicity. A total of 3.64 lakh people have benefited from the project.

The project has been successful in other ways, too. It has strengthened local institutions and allowed the beneficiaries to actually manage the reclamation activities. It has increased the income of marginal farmers and the landless, and given confidence and skills to the beneficiaries so that the results of the project are sustained over time.

ISRO's remote sensing satellites had a big role to play in the project. I travelled in Jaunpur district to find out how successful the project was and what was ISRO's contribution. Travelling with me were Dr A.K. Singh, the project scientist, and Dr M.S. Yadav, the project manager, both of the UP State Remote Sensing Centre, Lucknow, and now working on UPSLRP.

'Tell me all about the UPSLRP,' I asked Dr Singh.

Dr Singh explained, 'Sodic land has been a big problem here. We mapped the sodic land in UP using remote sensing satellite data in 1986–88. It showed 1.37 million hectares under sodic soil.'

'What exactly is sodic soil?' I asked.

'Sodic soil has high concentration of exchangeable sodium,' Dr Singh explained. 'So it has very low infiltration and hydraulic conductivity. This results in poor moisture transmission.'

'How would you explain that to a lay person like me?' I asked.

Dr Singh laughed as if I had said something very witty. 'That there is a high build-up of salt in these lands,' he explained. 'So high that water and air cannot penetrate.'

'How do you treat such land?' I asked.

Dr Singh said, 'Well, we have to replace the exchangeable sodium with calcium by addition of soluble calcium through external amendments. Gypsum and pyrite are most commonly used because they are easily available. Gypsum is the more popular amendment, because its reclaiming efficiency is higher than that of pyrite.'

'What that means is that you add gypsum to the sodic soil, is that it?' I suggested.

'Very simply put, yes,' Dr Singh conceded. 'The calcium from gypsum replaces sodium, leaving soluble sodium sulphate in water which is then leached out. This requires assured water supply or an irrigation system.'

'Why is that?' I asked.

'Because water is needed for leaching of salts from the root zone,' Dr Singh explained, 'for dissolving the gypsum to provide calcium ions to replace sodium, and finally, for crop production.'

'You should tell him about the requirement for drainage, Dr Singh,' Dr Yadav suggested.

'Yes, yes,' Dr Singh said. 'I was going to do that. Drainage of land is very important in reclaiming sodic soil. It improves the environment for crop growth and prevents build-up of salts by removing the excess water.'

'How do you do that?' I asked.

'We provide a surface drainage system to remove the excess rain and irrigation water,' Dr Singh explained. 'We've had to construct field drains to take out water from every plot of land to the main drains. That means, each field drain has to be connected to the main drain by an intermediary link drain. All the three types of drains—the field drains, the intermediary link drains and the main drain—have to be constructed before the process of reclamation starts.'

'That's a lot of work,' I pointed out.

'Yes', Dr Singh said. 'But that's where satellite remote sensing comes in. We have to scout around for natural drain lines, if they are available. In that case, the cost of constructing new drains, which, in any case, is the major component in sodic land reclamation, can be avoided.'

'Do you locate water by satellite remote sensing?' I asked.

Dr Singh nodded. 'Yes, that's an important part of mapping by remote sensing. We have to make sure that good quality groundwater is available at a moderate depth for pumping so that it can be used for reclamation.'

'Any requirement about the quality of water?' I asked.

'Yes, yes,' Dr Singh said. 'That is a must. The quality of the groundwater within the reclamation site has to be good. That's because irrigation in the programme is based on groundwater exploitation. We've to make sure that the salinity and sodicity values of the groundwater in terms of electrical conductivity and residual sodium carbonate are less than the prescribed minimum.'

'Tell me,' I asked. 'Didn't they try to reclaim these sodic soils in the past? I remember reading about it.'

'That's true,' it was Dr Yadav who intervened to point out. 'Uttar Pradesh has launched several schemes since 1945 to treat such lands.'

'How did these schemes fare?' I asked.

Dr Yadav made a face. 'Not very well, I'm afraid. That is because these schemes had a number of weaknesses.'

'Like what?' I asked.

'For one, they were not planned properly,' Dr Yadav explained. 'More importantly, there was no benchmark data to identify the sodic areas systematically. To add to that, there was inadequate understanding of the technology package, in the sense that attention was not paid to the need for drainage and assured irrigation.'

'What about the present scheme?' I asked. 'The UPSLRP, I mean?'

'UPSLRP takes care of all these aspects,' Dr Yadav said.

'It is very strong on the benchmark data for identifying the sodic areas. And also, strong on the data on location of drainage alignments and sources of assured water.'

'Is that where satellite imagery comes in?' I asked.

'Yes,' Dr Singh intervened to say. 'ISRO had already mapped sodic lands using remote sensing satellite data way back in 1986–88 under a national salt-affected soil mapping project. But this mapping was not good enough for the UPSLRP.'

'In what sense?' I asked.

'There was some classification problem,' Dr Singh explained. 'Some sodic lands are barren lying as waste while others grow poor crops. All these lands needed to be categorized properly. So what we do is to divide sodic lands into three categories. These are B+, which shows double-cropped land, B for single-cropped land and C for barren land.'

'How did you categorize them?' I asked.

'We had to depend on aerial photographs,' Dr Singh said. 'That is for delineation and categorization of sodic lands into B+, B and C types. We also depended on soil characteristics.'

'What were the other things you looked for while mapping the area?' I asked.

'Several things in fact,' Dr Singh said. 'Let me explain. There are two steps in using the satellite data for mapping. In the first step, we have to prepare the wasteland maps, distinguish the wasteland into saline or sodic and indicate the groundwater condition in general.'

'What about step two?' I asked.

'That is about the specifics,' Dr Singh explained. 'We have to map availability of large extent of sodic land in a contiguous area. Then we have to indicate the broad type of sodic land: barren, single or double cropped.'

'B+, B or C?' I suggested.

Dr Singh nodded in agreement. 'Precisely. In step two, using satellite data we have to map both groundwater depth and quality. In addition, we have to indicate general drainage conditions in the area.'

'What kind of a map did you generate, based on satellite imageries?' I asked.

'We did a fairly comprehensive wasteland map, superimposed with information on drains and streams, groundwater depth and quality, and coupled with ground truth data. This was for selecting the reclamation sites.'

'That's brilliant,' I pointed out.

'But these are only general maps based on the satellite data,' Dr Singh said. 'The complex part is to make them village specific.'

'Why do they have to be village specific?' I asked.

'Because that is what is needed for the project,' Dr Singh said. 'Village-wise reclamation plans had to be prepared. But the mapping for that had to be based on data from two sources other than ISRO satellite imageries. Aerial photography was one source. We also relied on cadastral data of the revenue department of the government. These were steps three and four.'

'What exactly was step three?' I asked.

'Step three was the detailed study in the villages within the selected reclamation sites,' Dr Singh explained. 'This meant village-wise mapping of the actual reclamation plots. We had to map the sodic land on a much higher but detailed scale, and characterize it in appropriate categories.'

'And step four?' I asked.

'That involved identification of plot number with the cadastral and ownership details,' he said. 'While doing this, we had to depend on the cadastral data provided by the UP government.'

'Why was that necessary?' I asked.

'Because,' Dr Singh explained, 'the land reclamation programme is to be actually implemented by farmers who own the sodic lands. So the ownership details of the sodic land plots are required.'

'How did you depict this aspect?' I asked.

He continued, 'We had to enlarge the aerial photographs with interpreted sodic land boundaries to the scale of the cadastral map, by matching permanent ground features and drawing the plot boundaries. You know something? In the map we finally prepared, all the sodic land plots could easily be identified.'

'That's brilliant,' I complimented him. 'But what about the ownership details?'

'We took care of that, too,' Dr Singh said. 'The ownership details of each sodic land plot had been collected from the revenue records. We put that in. So what we had in the maps was integrated information on plot-wise identification, characterization and categorization of sodic land in the form of maps, tables and ownership details for execution of the reclamation programme.'

Dr Singh looked triumphant. That was certainly a complex job very well done.

'That's really comprehensive,' I said. 'In other words, what you did by way of mapping was a multistage approach involving satellite data, aerial photographs and cadastral data.'

'Yes,' Dr Singh agreed. 'That, more or less, sums it up. Except with one difference. In the later stages, we had access to high resolution satellite data from ISRO's state-of-the-art remote sensing satellites. So we didn't really need the aerial photographs. We could do detailed sodic land mapping at cadastral level through interpretation of high resolution satellite data.'

'That's splendid,' I pointed out. 'Do you believe that the work you did contributed to the successful implementation of the sodic reclamation project?'

'In a sense, yes,' Dr Singh acknowledged rather modestly. 'Let us put it this way. What we did established a new approach to sodic soil reclamation. It was a crucial component in developing an effective model for environmental protection and also, improved agricultural production. Let me tell you that there were many other factors which contributed to the success of the reclamation project.'

'Like what?' I asked.

'I can list at least two. First, strengthening local institutions. Second, strong beneficiary participation in the project.'

I was incredulous. 'Did the beneficiaries actually participate in the implementation of the project?' I asked.

'They did,' Dr Singh asserted. 'And very competently, I must say. Beneficiaries were involved in decision-making from the earliest planning stage right through to the end of the project.'

'That's rather nice,' I pointed out. 'But who motivated them?'

It was Dr Yadav who intervened to say, 'The NGOs did. With their assistance, ten to fifteen beneficiaries were organized into units of 4-5 hectares of command area.'

'What did these units do?' I asked.

'They did a lot,' Dr Yadav explained. 'At the initial stages of the reclamation process, the units helped in verifying the details of the maps we had prepared. They provided the ground truth, and thus, the corroboration.'

'That would have been useful,' I agreed. 'Anything else?'

'Yes, yes,' Dr Yadav was quick to point out. 'With technical support, these units carried out all the site activities. Like tube well boring, field preparation and leaching. In fact, all the essential stages of the reclamation process.'

'What about funds, though?' I asked. 'All these activities need money, don't they? If it is a government programme, the funds would have been bureaucratically controlled. Not easy to get, I mean.'

Dr Yadav gave me a wry smile. 'Not really,' he said. 'The project funds were released directly into the accounts of these units of the beneficiaries.'

'How was the accounting done?' I asked. 'The government always makes heavy weather of accounting for public funds, doesn't it?'

'It was different with UPSLRP,' Dr Yadav asserted. 'The rules were different, and in a way, unusually flexible. Detailed accounting was done by these units of beneficiaries for funds released to them with the help of NGOs. What is interesting, and also different, is that once the sodic land was reclaimed, these units of beneficiaries were made responsible for the maintenance of the pump sets and drains linking their fields to the main drains.'

'That is the right thing to do,' I said. 'Dr Singh talked of village institutions. What were they?'

'There was another important institution at the village level,' Dr Yadav explained. 'That is the Site Implementation Committee. It consisted of all the beneficiaries at the village level.'

'What does the Site Implementation Committee do?' I asked.

'It allocates physical and financial resources to all the unit groups of water users,' Dr Yadav said. 'It also chooses the community resource persons. It resolves conflicts. It also monitors the implementation of the project.'

'Who are these community resource persons?' I asked. 'What do they do?'

Dr Yadav continued, 'They are the men and women who are selected by the Site Implementation Committee and trained to serve as voluntary change agents. They ensure that the beneficiaries understand the reclamation technology.'

'That's needed,' I pointed out.

'There is one more thing that the Site Implementation Committee does,' Dr Yadav explained. 'It also selects village animators.'

'What do these village animators do?' I asked.

'They serve as links between the villages and the technical social service agencies,' Dr Yadav said. 'Particularly so, after the project is completed. But on the whole, it is the NGO that helps in catalysing beneficiary participation in the project.'

'Didn't the project think of promoting self-help groups?' I asked.

'It did,' Dr Yadav pointed out. 'The project did moot the idea of self-help groups, particularly of women.'

'Did the idea work?' I asked.

'Eminently so,' Dr Yadav said emphatically. 'Self-help

groups of women became important hubs of economic activities in the sodic villages.'

'In what way?' I asked.

'In several ways,' Dr Yadav explained. 'They took up a wide range of micro-enterprises, supported by credit from banks and other financial institutions. They involved themselves in dairy farming, tailoring, growing nurseries for tree plantation and trading.'

'That's impressive,' I said. 'What about thrift and credit?'

'That too,' Dr Yadav said. 'The banks sanctioned cash credit limits to these self-help groups so that they can lend to their own members and to other villagers. Something very interesting emerged in the process.'

'What is that?' I asked

'That the repayment for the loans by the villagers has been exemplary,' Dr Yadav pointed out. 'This is something of a lesson to the banks in how to recover dues.'

'That's interesting,' I said. 'Is there anything else that these self-help groups did?'

'Yes,' Dr Yadav said. 'The groups conducted adult literacy programmes. On the whole, these self-help groups of women have been responsible for empowerment of women in the rural areas of Uttar Pradesh.'

'That's a significant achievement,' I agreed.

Dr Yadav had a smile on his face. As if he was about to tell me a joke. He said, 'Let me tell you that the success of these women's self-help groups has been so dramatic that self-help groups for men are beginning to form now.'

We were approaching our destination, Kurni village in Madiyahun block in Machhali Sahar tahsil. Dr Yadav had arranged for us to meet five beneficiaries of UPSLRP in that village.

※※※

The first beneficiary we met was Muni Raj. His total landholding consisted of 1 acre of land which had been affected by sodicity, and had been classified as totally barren.

'May we ask you a few questions about the Sodic Land Reclamation Project?' I asked Muni Raj.

'Of course, you can,' he said flashing me a smile.

'When did the implementation of the project start in your land?' I asked.

He scratched his head. 'I don't exactly remember,' he said, but he looked thoughtful. 'Well, let me see. That was the year my niece gave birth to her first son. He is about four years old now. So it was four years back.'

'What did you grow in your land before the project started?' I asked.

He shook his head. 'Nothing,' he said. 'Absolutely nothing. It was lying barren for several years at a stretch.'

'Why was that?' I asked.

'Because nothing would grow. I tried several times. But looking at how miserable the crops were, I finally gave up.'

'What happened when the project started?' I asked.

'Well,' he said slowly, 'we did several things. Things like drilling a tube well, preparing the field, leaching the salt out and then taking a fresh crop.'

'You said, "we",' I asked, 'didn't you? Were you really involved in all that work?'

He nodded his head emphatically. 'Of course, I was,' he said. 'Do you mean to say I shouldn't have been?'

I pointed out quickly, 'No, no, I didn't mean that. I was only trying to find out how far you were involved in the whole thing.'

He twirled his moustache. 'I was even a member of the water-user group,' he said triumphantly.

'Now that your land is reclaimed, what crops do you grow?' I asked.

'I grow two crops now,' Muni Raj said. 'I grow paddy as the kharif crop. And wheat as the rabi crop.'

'That's great,' I said. 'What is the productivity like?'

Proudly he said, 'You won't believe it, but I get something like 6 quintals of paddy from my land and 5 quintals of wheat.'

'How does it compare with the productivity of your land before it was reclaimed?' I asked.

He spat repeatedly on the ground as if to say that he would not like to be reminded of those harrowing days. 'Well,' he said finally, 'since you are asking me, I must tell you. In those days, I got nothing from my land. Zero. Let me tell you that whatever I get now is a bonus.'

'What did you do when the land gave you nothing?' I asked.

He spat again. 'Those were hard days,' he said. 'My entire family had to work as agricultural labour.'

'Does that mean that your income has increased considerably?' I inquired.

'Certainly, yes,' Muni Raj said. 'I forgot to tell you something. I also grow daincha in my land, but that is only a pre-kharif crop.'

'What exactly is daincha?' I asked Dr Yadav.

'Oh, that's used as green manure for the kharif crop,' Dr Yadav explained.

'Did you also get trained in improved agricultural practices?' I asked Muni Raj.

'I got a lot of training,' Muni Raj said. 'And let me tell you that it was very useful. For example, that's where I learnt about how to grow and use daincha.'

'Did you learn other things, too?' I asked.

'Yes,' Muni Raj asserted. 'A lot. They told us about the seeds and fertilizers and how to use them. I think the most important thing I learned was how to use water. Tell me something, isn't the yield I get from my land good now?'

'Yes,' I assured him. 'What you get by way of paddy and wheat yield sounds very good.'

'I thought so,' Muni Raj said. 'In my case, I can't tell you what a blessing it is to have good crops growing in your land. And just to think that only a few years back, I got nothing from it. Nothing at all!'

'You must be a relieved man,' I suggested.

'Yes, I am,' he said with a sigh. 'It's not that I've totally given up doing wage work; I do it sometimes, only to supplement my income from the land. But in the old times when my land gave nothing, I had to be a wage labourer throughout the year. I even migrated to the town in search of work and stayed there.'

'What happened to your family?' I asked.

'That was the problem,' Muni Raj said. 'I didn't see them for months on end. Now, with my land giving me such a lot, I am a contented man.'

※※※

The next farmer we met was Babu Ram. In terms of landholding, he was better off than Muni Raj because he owned three acres of land out of which 2 acres were sodic land. But those 2 acres were totally barren and classified in C-category.

Babu Ram was happy to see us. He said, 'When Dr Yadav told me that you people were coming to talk about this project, I thought it was a good thing.'

'Why so?' I asked.

'Because I wanted to say what this project has done for us,' he said. 'It has done a lot of good for me.'

'In what way?' I asked.

'In so many ways,' Babu Ram said cheerfully. 'It has given us productive land in place of barren land. To tell you the truth, it has put food back in the mouths of my family.'

'Tell me, what used to happen before the project started?' I asked.

'Terrible things,' he said with a grimace. 'I tried everything. Rice and wheat wouldn't grow. I even put mustard. That too met the same fate; I tried that for several years, but without luck. It gave me such poor crop that even our cattle refused to eat it.'

'So what did you do?' I asked.

'Luckily for me,' Babu Ram continued, 'I have this one acre of non-sodic land. I grow sugar cane there. That gives me some income, but not enough to keep my family going for the entire year. So we had to do wage labour to keep our body and soul together. That's a lot of hardship, I can tell you that.'

'Have things improved after the project?' I asked.

'Oh, yes!' His voice was emphatic. 'I now grow paddy, wheat, mustard and daincha in these two acres. Let me tell you I make enough now to meet our hunger.'

'What's the kind of yield do you get now?' I asked.

'I get plenty,' Babu Ram said with glee. 'I get about 18 quintals of paddy per acre. For wheat, I get about 17 quintals per acre.'

'That's good yield,' I agreed.

'Isn't it?' Babu Ram said. 'And just to think, nothing would grow in that land before. Not even a blade of grass!'

'Did you participate in the project?' I asked. 'What I mean is, were you involved in the actual implementation?'

'Yes, I was,' Babu Ram said with a tinge of pride in his voice. 'Do you know something? I was a member of the Site Implementation Committee.'

'What does the Site Implementation Committee do?' I asked.

'All kinds of things,' Babu Ram said, waving his hands in the air. 'It gives money for all the works in the project like the tube well, drainage works, preparation of the field, etc. When people fight, it is our committee that settles the fights. We also keep an eye on how the project is going.'

'That's an impressive range of functions.' I pointed out. 'But tell me, to what extent were you involved in the planning of the project?'

'I was there at every stage, participating in every little bit of work. Let me tell you something. When those big coloured maps came, showing our lands, they asked me whether my land was there. What do you call those coloured pictures?' He asked Dr Yadav.

'Those are our remote sensing satellite imageries with cadastral inlay,' Dr Yadav explained.

Babu Ram nodded. 'Yes, that's what they are. They

wanted me to verify what was shown there. Also, the information about my land. I told them what was right and what was wrong.'

'Were you also involved in the actual works?' I asked.

'Yes, of course,' he said. 'Right from the boring of the bore well to the leaching of the salt. We were the people who did all the work. Of course, the engineers and others were there to tell us what to do.'

'What about training?' I asked.

'Yes,' he asserted. 'We were given training. That's how I learnt all the things about how to grow a good crop. What do you call these things in your language?' He asked Dr Yadav.

'Improved agricultural practices,' Dr Yadav explained.

Babu Ram nodded. 'Yes, that's what it is,' he said. 'They told what seeds to plant, where to get them from, how much time the crop will take to grow if we used those seeds. They even have a name for it. What is that?' He asked Dr Yadav again.

'Certified Seeds,' Dr Yadav explained.

'Yes, yes,' Babu Ram said with a nod. 'They also told us about fertilizers. What kind to use and how much to put. They also told us about how to use the water from the tube well for our crops. Not too much, not too little. Just the right amount.'

'Did all that advice work?' I asked.

'Of course, it did,' Babu Ram said with certitude. 'How else do you think I could have got 18 quintals of paddy and 17 quintals wheat from one acre of land? Let me tell you how much mustard I got. My house had never seen so much of mustard before. And all this from that 2 acres of land which refused to give me anything before, not even a blasted blade of grass!'

'That's really impressive,' I pointed out. 'You mentioned something about green manure, didn't you?'

Babu Ram nodded. 'Yes, I did,' he said. 'I now grow daincha. I use it as manure for my crops. This I came to know only when they gave us our training. Just to think that I didn't know about something so useful which you can grow easily!'

'Have you benefited from this project?' I asked.

'Yes, I have,' Babu Ram said. 'To tell you the truth, this has been a godsend for my family members. I am now in a position to spend more and more time with them. Because I am there at home, I can keep an eye on my children and see that they study and go to school regularly. I honestly think this project has been a real boon to farmers like me.'

※※※

Suita Ram was the farmer we met next. Like Babu Ram, he has three acres of land, but half of his landholding was affected by sodicity.

'You are the richest of the farmers we have met so far,' I said to Suita Ram.

He knit his brow in puzzlement. 'Is that so?' he said diffidently. 'I don't think I can be called rich in any way. I am just a farmer who makes both ends meet with some difficulty.'

'I said so because only half of your lands happened to be sodic, while in other cases it was more,' I pointed out.

He looked relieved.

'Oh, that way,' Suita Ram said. 'But let me tell you that the one-and-half-acres of my land which were sodic, were absolutely barren. Nothing would grow there.'

'Did you try to grow something in that land,' I asked.

He shrugged. 'Oh, yes!' he said stoically. 'Many times. I would plant something year after year. I tried so many crops: potato, wheat, paddy, vegetables. Nothing would grow!'

'When did the reclamation of your land start?' I asked.

'Four years back,' he said. 'That's the time the planning started. Tube wells were bored. Drains were built and they leached the salt out of my land.'

'Did you take part in those activities?' I asked.

'Yes, I did,' he said enthusiastically. 'In fact, I was a very active member of the water user group.'

'Were you in the Site Implementation Committee?' I asked.

'Yes,' Suita Ram said. 'I was a member of that committee.'

'What kind of crops do you grow now in the reclaimed land?' I asked.

'Oh, all kinds!' he said, 'potato, sugar cane, wheat, paddy and vegetables.'

'That's a lot,' I said. 'Do these crops grow well in the reclaimed soil.'

He nodded vigorously. 'Oh, yes,' he said. 'Like in any other normal land. Let me tell you how much wheat I got from the last crop I took. I got about 18 quintals of wheat from half an acre.'

'But that seems to be even more productive than the normal land!' I pointed out.

'Maybe,' Suita Ram said. 'That's because we know now how to get a good crop. Right kind of seed. Right amount of good fertilizers. And of course, the right quantity of water at the right time. This is all thanks to the training we were given in the project.'

'That's very nice,' I said. 'Is there any other way that you benefited from the project?'

Suita Ram thought for a moment and then said, 'I think the greatest benefit I've got from the project is the food that it has given me and my family. What I get from my lands now meets the entire food requirement of my family and also something more.'

'Do you mean that you have surplus food left after your family's consumption?' I asked.

'Yes, that's true,' he said delightedly. 'I do have a surplus now.'

'What do you do with the surplus,' I asked.

'I sell it in the nearby market and make some money,' he said.

'That's great,' I suggested. 'And what exactly do you do with that money?'

'Several things,' he said proudly. 'Useful things, of course.

I have used a part of that money to instal a sugar cane crusher. That has been a good investment because that also brings me some income.'

'Anything else?' I asked.

'Yes, yes,' he said. 'I have bought some livestock. I hope to make some money out of that, too.'

'That's clever thinking,' I pointed out.

'Isn't it?' Suita Ram said gloatingly. 'I am also thinking of buying some additional land for cultivation.'

'I'm impressed,' I said. 'Could you have done all this if your sodic land hadn't been reclaimed?'

Suita Ram shook his head. 'Not a chance!' he said. 'That time, when half of my lands were lying barren for years together, I was in great difficulty. I couldn't support my family from what I made from my lands. Let me tell you something. I was heavily in debt. Thank God, I've paid back those debts and I've a little money now. This project has brought Lakshmi to my house.'

※※※

The next person we met in Kurni village was Ram Shankar. He has two acres of land out of which one acre was sodic. The sodic land in Ram Shankar's holding was classified as C-class: it was barren.

'Was your land reclaimed under the project?' I asked Ram Shankar.

'Yes,' he said. 'That was for one acre of my land which was affected by salt.'

'What did you grow in that land before the project started?' I asked.

Ram Shankar shook his head in desperation. 'I tried very hard,' he said plaintively. 'Every year I would plant something. In the early years, some crop would come up, but so poor that the yield was almost nothing.'

'What crop did you plant?' I asked.

'I tried everything,' Ram Shankar said. 'Wheat, paddy, tomato, potato. But nothing succeeded. So I finally gave up.'

'Now that the land has been treated,' I said, 'what do you grow?'

'You'll be surprised,' he said with a twinkle in his eye. 'The very same crops I tried before. But the growth is so luxuriant now. It's like any other normal land.'

'What's the yield like?' I asked.

'Very good,' he said. 'I take a good potato and tomato crop. I've tried paddy and wheat. In the last crop, I got about 12 quintals of paddy. In the crop before, I got 10 quintals of wheat.'

'That sounds like a pretty good yield,' I suggested.

'Yes,' Ram Shankar said. 'It can be better, though. But I'm happy. If I can get so much from a plot of land which gave me nothing for years together, it's all for the good.'

'Did you participate actively in the project?' I asked.

'Yes, I did,' he said. 'It was a good project. They involved us at every stage. Almost everyone whose land was being reclaimed.'

'Would you say that the project was managed by the beneficiaries,' I asked. 'By farmers whose plots of land were being treated?'

'I think so,' Ram Shankar said. 'The beneficiaries did everything. But of course, the guidance, particularly in technical matters, came from the experts.'

'That's how it should be,' I said. 'Tell me, how did you benefit from the project?'

Ram Shankar pondered my question for a minute and then said, 'In several ways. It was the training which was most useful. It taught me so many things about how to raise a good crop.'

'Anything else?' I asked.

'Yes,' he said. 'It was the forming of the self-help groups under the project. If you ask me, this was the best thing that the project did for us next to reclaiming our sodic lands.'

'Why do you say that?' I asked.

'Because the self-help group is such a good idea,' he said. 'And it has done so much for poor people like us. It was the project that made our women form these groups.'

'Were you involved with these groups?' I asked.

'No,' he said, 'but my wife was. You know what these groups did? They started by collecting contributions from the members and rotating the money to finance our credit needs.'

'Was that all?' I asked.

'No,' he said, 'that was what they started with. But they did very good things. They gave money for all kind of things.'

'Like what?' I asked.

'Many things,' Ram Shankar explained. 'Tailoring, grocery shops, trading activities, dairying, milk business. Do you know something? These groups even gave money for raising nurseries for tree plantation.'

'That's very laudable,' I pointed out. 'Tell me, where did these groups find so much of money to fund such activities?'

'Well,' he said, 'they went to the banks and the banks gave them the money. Everyone said that is money down the drain. Poor people never pay back. But you should've seen what the recovery was like!'

'How was it?' I asked.

'Fantastic,' he said gloatingly. 'Hardly any arrears. That only shows that poor people can be trusted to pay back loans.'

'Obviously,' I said. 'Tell me did the groups do anything else apart from thrift and credit?'

'Yes,' he said, 'they conducted adult literacy classes. I went to these classes myself. I can tell you that I learnt such a lot from them. Let me tell you that other illiterate villagers here in Kurni village learnt a lot from these classes, too.'

'That's so creditable,' I said. 'I don't understand one thing, though. Where did the illiterate women of your village get the strength to do so many things?'

'That amazes me sometimes,' Ram Shankar said. 'Obviously it was the reclamation project. Our women got the strength from that. This project has done so many things for us. It has really helped us.'

※※※

The last person we met in Kurni village was Ram Khelawan. He has a landholding of two acres out of which one acre was sodic land classified as C.

'So you had one acre of barren land, afflicted by salt,' I asked Ram Khelawan. 'Isn't it?'

'That's true,' he said with a shy smile. 'Let me tell you that one acre of land gave me no end of trouble!'

'Why do you say that?' I asked.

'Endless trouble,' he persisted. 'We were driven mad by that patch of land. We would go on planting all kinds of crops but nothing would grow. We tried wheat. We tried paddy. Even bajra!'

'Why do you say "even bajra"?' I asked.

'Because,' Ram Khelawan said, 'they say here in our villages that bajra can grow anywhere. Even on a rock! But that piece of my land was so wretched that even bajra plants wouldn't sprout. Can you believe it?'

'These sodic lands can be difficult to cultivate,' I said. 'Tell me, now that your land is treated, is it any different?'

'Of course, it is!' Ram Khelawan crowed. 'It has become like any other piece of fertile land. Do you know what I planted after it was reclaimed? Paddy and wheat. And bajra.'

'How was the bajra crop?' I asked.

There was a gleam in his eyes. 'Magnificent,' he said. 'I have never seen bajra growing so luxuriantly anywhere else.'

Ram Khelawan had scored his point with the land, I thought.

'Tell me,' I said, 'how was the yield?'

'Fantastic!' he said. 'I got something like 12 quintals of paddy per acre and 10 quintals of wheat.'

'Was that enough to meet the food requirements of your family?' I asked.

'More than enough!' he said gloatingly. 'I've a small family, and what I get from my lands after the reclamation is adequate. This is the first time that we can meet our food requirement from our lands.'

'The way you say it,' I suggested, 'means that it is something special for you, isn't it?'

'Yes,' Ram Khelawan said, his voice surging, 'it is something special. Do you know what used to happen? Before our sodic land got treated, we had to purchase food grains from the market. And it is such a disgrace!'

'A disgrace?' I was incredulous. 'Why should it be? I do it all the time. I buy my food grains from the market.'

Ram Khelawan laughed. 'But you are a town man,' he said. 'You have to buy your food grains from the market. You have no choice. But in the rural areas, if you buy your food grains, it is socially downgrading.'

So that's what the project had done for these people: by improving their food security, it had given them social esteem. Not a mean feat, by any standard.

'I agree,' I said. 'The project has given you social prestige. Has it done something else for you?'

There was no hesitation in his voice as he answered my question. He said, 'Let me tell you about something that really happened to me. There is this man in our village who owns vast tracts of lands and is certainly the wealthiest person in these parts.'

'A political leader?' I asked.

'Sort of,' Ram Khelawan said. 'When this sodic land was troubling me and I was heavily into debts, I approached him. I asked him to buy that sodic land.'

'What did he say?' I asked.

'He laughed in my face,' Ram Khelawan said. 'Laughter of derision, if you please. He said, "That's a piece of junk. Who'll buy that? I won't even take it as gift!" He shooed me away as if I were a mad dog!'

'Why would he say that?' I asked.

'Because that was the truth,' Ram Khelawan said. 'That sodic land was a piece of junk. Let me tell you what happened. He came to my house the other day.'

'Did he?' I asked. 'What did he say?'

'He offered to buy the same piece of land,' Ram Khelawan said. 'He offered me a price that was incredibly high. I said a very firm no. How can I sell a piece of land that is like a pot of gold? Did I do a wrong thing?'

'No, no,' it was my turn to say. 'You did the right thing.'

Dusk was setting in. Cows were coming home, kicking the dust off the village road. It was time for us to leave. As we said our goodbyes to Ram Khelawan, what he had said about his reclaimed land being a pot of gold kept ringing in my ears. The sodic land reclamation project in Uttar Pradesh had turned unproductive, barren soil into fertile, arable land, or as Ram Khelawan put it so aptly, into pots of gold.

12

IN GOD'S OWN COUNTRY
Automatic Weather Station

Madhavan Nair, the present head of ISRO, has distaste for formality. The distaste goes beyond wearing bush shirts on all occasions to the way he approaches the complicated task of building a rocket. He likes to derive things from first principles, often drawing simple pictures of very intricate structures, much like the physicist, Richard Feynman, who drew simple pictures to describe the dynamics of fundamental particles.

It is from such simple pictures that Madhavan Nair built the PSLV, the launch vehicle which is now the workhorse of ISRO to launch remote sensing satellites. With a success rate of seven out of eight, PSLV is that rare vehicle that makes insurance companies stay in business. PSLV is a good-looking vehicle, too. Madhavan Nair obviously has a sense of aesthetics; qualities rather rare in engineers who build huge, complex structures.

It is also to such simple pictures drawn by Madhavan Nair that the Automatic Weather Station (AWS) owes its birth. The AWS records weather data such as temperature, atmospheric pressure, rainfall, wind speed and direction, relative humidity and solar radiation on a continuous basis.

It is a compact, modular, rugged low-cost system which is capable of operating with minimum power from battery and solar panel for extended periods in the field in remote areas.

I went across to Madhavan Nair's room to talk to him about the AWS.

'How did you think of the AWS?' I asked him.

'Well, the idea was with me for quite some time,' he explained. 'It is because I wanted to do something for agriculture to benefit people living in villages.'

'But how does weather come into it?'

'That's because agricultural operations are weather dependent. Operations like ground preparation, tilling, sowing, weeding, fertilizer or pesticide application, irrigation and harvesting are decided on the basis of the weather situation and trends. Important decisions like the selection of the crop depend on the arrival of the monsoon and its expected performance. And post-harvest operations like drying and transportation also depend on the weather. They need fair weather.'

'That's true,' I said.

'Well, the agricultural operations we talked about need information on current weather. But the planning and management of agricultural operations require forecast of weather over the next few days.'

'And also an outlook for a few weeks ahead,' I suggested.

'Yes. But let me tell you that the agriculture sector needs something else. There are severe weather events such as floods, cyclones, frost, cloudburst or even drought. The agriculture sector requires warnings on these severe weather events. In addition, the insurance coverage to the agriculture sector at the farm level depends on weather information with adequate spatial coverage and frequency.'

'Will the AWS improve productivity in the agriculture sector?'

'Oh yes. What the AWS does is to provide weather information in the villages on current weather situation. This reduces the risk involved in agricultural operations and leads to improved productivity.'

'But what are the kinds of things that would influence weather?'

'Well, let me tell you something to start with. There's a very delicate balance between land, ocean and atmosphere. As a result, the interactive process between these three is very subtle. That makes the Earth's system a non-linear, coupled system.'

'Are you suggesting that weather is difficult to predict?'

'Yes. That is when one tries to predict weather by simple numerical methods. Let me give you an example. The local weather is affected by atmospheric conditions, no doubt. It's also affected by regional and global systems. Therefore, it is necessary that we observe the atmospheric conditions over the entire globe continuously.'

'What are the time scales for weather forecasting?'

'It can be short, which is about twenty-four hours. It can be medium, which is about three to five days. Or, it can be seasonal which is monthly.'

'Do you have weather models for making these forecasts?'

'Yes. They are based on fundamental principles governing atmospheric flow. They simulate the likely condition of the atmosphere in various time scales.'

'What are the inputs to these models?'

'Inputs are the accurate observations of the atmospheric parameters such as pressure, temperature, wind, humidity and radiation.'

'It is often said that making forecasts here in India is much more difficult, than, say Europe. Is that true?'

He nodded. 'Yes. That's because the relationship between the forcing functions and atmospheric processes is more linear there. In tropical regions like India, the processes are more complex, leading to more non-linearity. Therefore, more efforts are required in terms of extensive observations and modelling.'

'Do you mean to say that there should be more observation points?'

'Yes. The observation network has to be expanded. Let me give you an example. Ideally, the observation network over the tropics should be made dense with one station every 100 square kilometres. At present in India, we have only 600 surface observations, thirty upper air observations, and a handful of radar stations. The spread is inadequate.'

'Is that how the AWSs will help in improving the observation network?'

He nodded. 'Yes. One of the key elements in improving the weather information services is the adequacy of weather observation, especially from local levels. That's the reason why we thought of having the AWSs and locating them at local levels, particularly in areas which are remote and inaccessible.'

'How exactly are you planning the location of these AWSs?'

'The idea is to set up the AWS at the Village Resource Centre and provide weather-related services to the rural community. And to develop suitable interface with the Indian Meteorological Department for generating the required services.'

'Did you develop the AWS in ISRO itself?'

'Yes. We developed a state-of-the-art AWS in collaboration with the Indian industry.'

'Is it an indigenous system?' I asked. 'Or does it have imported components?'

'Totally indigenous. And state-of-the-art, too. The pressure sensor is built at ISRO. For the other systems, we've passed on the technology to the industry.'

'Do you have such systems abroad?'

'Yes. In fact, one of the multinational companies markets such systems in India. But let me tell you, that the one we've developed is as good, if not better. But I think, the big difference between our AWS and theirs is the cost.'

'What is the cost differential?'

Madhavan Nair's voice was brimming with pride as he said, 'We give it at one-fourth of the price as compared to the multinational company.'

※※※

Dr Manikiam and I were travelling to Wayanad district of Kerala to find out how the users could benefit from the AWS. Dr Manikiam is the weather expert at ISRO. He has contributed greatly to the development and building of the AWS. We were driving from Mysore to Sulthan Bathery. I started talking to Dr Manikiam about the AWS.

'Tell me,' I asked. 'What does this AWS consist of?'

'It consists of weather sensors,' Dr Manikiam explained. 'They measure various parameters such as temperature, pressure, wind velocity, humidity, radiation, rainfall, etc. There is also a datalogger in the AWS unit here.'

'What does the datalogger do?'

'It receives the data from the sensors and converts them to digital signals. Then, it sends the signals to ISRO's satellites.'

'Which are these satellites?'

'We have INSAT-3A as well as Kalpana. Kalpana is a dedicated meteorological satellite. There is a Data Relay Transponder onboard these satellites. DRT for short.

'What does the DRT do?'

'It receives signals from the AWS and retransmits them to the ground receivers. At the receiving end, the data stream is again converted to retrieve the sensor observations.'

'Have you stipulated a time when the AWS should transmit signals?'

'Yes. Each AWS transmits at a particular, previously decided time twice a day. This helps in avoiding data loss due to simultaneous transmission by AWS.'

'What are the other capabilities of the AWS?'

'It has several capabilities. Like easy programming of the sensors, front panel display, archival of one year's data.'

'How many such AWSs can ISRO's satellites handle?'

'We can handle several thousands of AWSs with both INSAT-3A and Kalpana in orbit.'

'What can be the possible applications of AWS data?'

'Primarily, it's the weather. It provides information on current weather as well as an assessment of the anomaly from normal or expected weather. There is another important use for the data generated by the AWS. You see, there is this National Centre for Medium Range Weather Forecasting. NCMRWF for short. This organization does modelling for weather and prepares forecasts. The forecasts generated by NCMRWF provides medium-range weather forecast of rainfall, temperature and winds for a duration of three to five days.'

'What about the agriculturists?' I asked. 'Do they use the medium-range weather forecast?'

'Yes, some agriculturists do use them. You see, there are these Agro Advisory Service units. This service is highly beneficial to the agriculture sector. It can be further expanded with the data from the AWS network.'

'Are there other uses for AWS data?'

'Yes. The AWS can be used for applications such as flood-level monitoring, pollution monitoring, irrigation scheduling, water management, etc. In fact, it can also be used for business operations, such as tourism.'

'Tell me something,' I asked. 'Can the AWS data be used by private organizations?'

'Yes. Specific advisory services related to agriculture like irrigation scheduling, fertilizer application, spraying of pesticides can be generated by agronomists and meteorologists, using data from AWS.'

'That, I think, would be a very good idea,' I suggested. 'Tell me, have such things worked in other countries?'

'Yes, they have worked quite well. You see, there are studies of mesoscale network of AWS stations in other countries. These are networks of 5–10 square kilometres grid. Such studies have shown improvement in local weather forecasts as required for cities, towns, airports and ports. The usefulness of the AWS has been proved worldwide.'

'What kind of services are you planning to provide through the AWS?'

'We are planning to provide current weather information in respect of air temperature in terms of maximum and minimum, surface winds, humidity, cloudiness, soil moisture, solar radiation, rainfall conditions, humidity and temperature.'

'This information will be provided to the farmers?'

'Yes. In addition, we'll provide updated satellite images—both visible and water vapour—through the website of the Indian Meteorological Department and the Space Application Centre, Ahmedabad. And specific derived parameters such as cloud cover and radiation.'

'What about weather forecasts?'

'Yes, that'll be provided. The medium-range weather forecast—of three to five days—given by NCMRWF will be extended to the locations of the Village Resource Centre. You see, the forecast generated by NCMRWF runs in a nested mode. It gives the forecast of rainfall, temperature and winds for three to five days.'

'What about the daily weather reports?'

'Yes, that will also be provided. That will be the daily weather report and short-range forecast—which is for twenty-four hours—issued by the Indian Meteorological Department based on synoptic weather data analysis and statistical models. Well, the daily weather reports of the Indian Meteorological Department include the progress of monsoon.'

'What about the agricultural advisory services you were talking about?'

'We're planning to issue specific advisory services related to agriculture such as irrigation scheduling, fertilizer and pesticide application prepared by agronomists and meteorologists.'

'Well, you'll be generating a whole host of weather data,' I suggested. 'Aren't you planning to use them for other purposes?'

'Yes, we've plans for that. You see, the weather data collected in a region or district would be useful to prepare analysis of things such as rainfall distribution. That could be block-wise or for a cluster of villages. Such analysis would help in indicating the anomaly in rainfall distribution.'

'What you mean is the deviation from the normal pattern, is that it?'

'Yes, yes. The weather data would also help in preparing

the cloud cover and moisture status for use in pest and disease surveillance models. The data will also be useful in compiling the soil moisture status, both computed and measured, as well as, moisture adequacy for agriculture. This would include crop-wise requirements.'

'Can't you use this data for water management?'

'We plan to do that. I've already told you about rainfall distribution and deviation from normal. The weather data would be useful to prepare analysis of the rainfall run-off and assessment of the groundwater recharge. It would also help in making drought assessment at the village level. This would make possible an analysis of incidence of drought.'

'What about public services?'

'Yes. The weather data will have potential applications in several areas of public service. Such as heavy rainfall and flooding, and climate data for planning developmental activities. And in the area of healthcare, when it comes to extreme weather.'

'How are you planning to give these weather-related services through the Village Resource Centres?'

'In a phased manner. Initially, the current weather and short-range forecast issued by the Indian Meteorology Department will be made available. Further, we plan to expand the service by extending the medium-range forecasts generated by NCMRWF to the locations of the Village Resource Centres after due validation. The weather data generated by the Village Resource Centre network will be systematically archived and made available for sectors such as watershed development, water management and insurance.'

'By the way, who are we meeting during our visit?'

'We are now going to meet a tourist operator in Sulthan Bathery. Didn't I tell you that the AWS would be useful for business operations? I suggest you ask him how he is going to use the information generated by the AWS.'

'I thought we were meeting agriculturists.'

'Yes, we are meeting them, too. That we'll be doing at Meppadi and Lakkidi. In fact, we are about to reach Sulthan Bathery.'

'Tell me,' I asked. 'Why is it called Sulthan Bathery?'

'You see, once upon a time it was known as Sulthan's Battery. That was because it was associated with Hyder Ali and Tipu Sultan. But today, Sulthan Bathery doesn't even have a fort!'

※※※

We went to the place in Sulthan Bathery where ISRO's Automatic Weather Station had been installed. It was a smaller structure than I had thought it was, but it looked compact and rugged. We went to see Ahmed Moosa, a resident of Sulthan Bathery.

Ahmed Moosa wore a kaile mundu around his waist and a thin kurta type shirt. The kaile mundu was tied with a belt—the Malabar belt—with a pouch obviously meant for keeping money. The belt also had a dagger hole, but Ahmed Moosa wore no dagger.

'What is your profession?' I asked.

'I'm a tourist operator,' Ahmed Moosa explained.

'What exactly does that mean?' I asked.

'Well, I arrange conducted tours for tourists. I also organize treks and expeditions for adventure sports as well.'

'That's nice,' I said. 'But tell me, is tourism picking up in Wayanad district?'

'Yes, yes. You see, Wayanad has now been successfully projected as the green paradise. The hills, rocks and valleys here provide great adventure experiences. The mountains and forests provide many trails, trekking and opportunities for adventure sports. Well, there are still vast areas in Wayanad waiting to be explored. That makes Wayanad an adventurescape waiting to be discovered. So we say: come, explore Wayanad.'

I beamed a smile at him. 'You are a good salesman. You have sold Wayanad to me. Maybe I'll come one day and explore Wayanad. Tell me, are things happening here?'

'Yes, things are happening. There has been a lot of

development in the last three years. Transport has improved. There is good advertising by the government. More tourists are visiting nowadays. Hotel accommodation has improved.'

'Tell me,' Dr Manikiam asked. 'What kind of tourists do you get here in Wayanad?'

'Well, most of them are from the neighbouring states. Some of them are from north India. But let me tell you that nowadays, there are tourists from USA and Europe who've started coming.'

'How do you plan these tours?' I asked.

'We are having links with booking agencies from all over India. They tell us the details. We do the local planning here.'

'What kind of weather information do you want for your business?' Dr Manikiam asked.

'Weather is very important for us to plan properly. If I can know weather conditions, I can do my business activities properly.'

'In what way?' Dr Manikiam asked.

'Let me give you an example. If I know that it's going to be a rainy day, I'll plan for my tourists to visit temples, museums. If it's going to be a sunny day, I'll plan trekking, boating, etc. There are many options.'

'What about the foreign tourists?' I asked.

'The weather information is particularly important for them. Do you know something? Many foreign tourists want to know in advance what type of weather it is going to be on a particular day.'

Dr Manikiam intervened to say, 'Suppose we give you daily weather information and the trend for the next three to five days, will it be useful? ISRO has established this AWS at Sulthan Bathery. It'll give you a daily weather report. It'll also give you a forecast of rainfall, temperature and winds for the next three to five days. Will that be useful?'

'No doubt. In that case, we can plan hour-to-hour programme. We can also brief our tourists properly.'

'Will that improve your business prospects?' I asked.

'Most certainly. If you can give weather forecasts three

days in advance, my business will certainly improve. Let me tell you something. We don't want our tourists to be stranded due to heavy rain, flooding and landslide.'

We took leave of Ahmed Moosa.

※※※

Our next destination was Meppadi. We were supposed to see the AWS there and meet Shri Raman Kutty. As we travelled to Meppadi, we started climbing the Western Ghats. And what a sight it was! There was a panorama of undulating low hills converted into paddy fields. The hills were full of plantations like tea, coffee, pepper and cardamom, while the valleys had a predominance of paddy fields.

'You're so engrossed in the scenery around!' Dr Manikiam told me. 'You find it beautiful, don't you?'

'Yes,' I admitted. 'It is incredibly beautiful. And it is so intensely green! Even the paddy fields here look greener than anywhere else!'

'The paddy fields,' Dr Manikiam pointed out, 'always engross visitors to Wayanad. Do you know what the name Wayanad means? It means Vayal Nadu, or the village of the paddy fields.'

'How appropriate,' I said.

After Kalpetta, we climbed further into the Western Ghats. The landscape was beautiful with lofty ridges, interspersed with magnificent forests, tangled jungles and deep valleys.

'Tell me,' I asked Dr Manikiam. 'Wayanad district is primarily agricultural, isn't it?'

He nodded. 'Yes, it has an agricultural economy. There are no major industries to boast of in this district.'

'What kind of crops do they grow here in Wayanad district?' I asked.

'Wayanad is situated at a height of 700–2100 metres above the sea level. So because of the high altitude, there is cultivation of perennial plantation crops and spices.'

'That's true,' I said. 'I do see a large number of coffee plantations.'

Dr Manikiam nodded. 'Yes, coffee is ubiquitous in Wayanad. You know something? It is cultivated in each and every panchayat here. In plantations as well as in small holdings. It is grown both as a pure crop and as a mixed crop along with pepper.'

'The famous Pulpally pepper?' I asked. 'Is that from Wayanad?'

'Yes, yes. You see, there are exclusive pepper gardens in Pulpally area. But otherwise pepper is grown as an additional crop to give shade to the coffee shrubs.'

'What are the other plantation crops in Wayanad?'

'There is rubber, coconut, cardamom, tea, cassava and ginger. Of late, they also grow vanilla.'

'What about paddy?'

'It is a major crop in Wayanad. The rice fields of Wayanad are in valleys formed by hillocks. But in most of the paddy lands, only a single crop is harvested.'

'I see a lot of ginger,' I pointed out.

'That's right. Ginger cultivation in Wayanad has increased substantially in recent times. You see, the ginger produced here is mainly marketed in the form of green ginger.'

I looked out of the car window. The landscape was very green. All around us were fields shimmering with a kind of green I had not seen before. It was an intense green everywhere; there must be something special about the soil of Wayanad, I thought, that promotes such luxuriant growth of vegetation which clothes the landscape in such uniform greenery.

Dr Manikiam broke into my thoughts. 'The special feature of Wayanad,' he said, 'is that, here, agriculture is done by homestead farming. But let me tell you that the average size of holdings here is very small.'

'How small?'

'Almost half a hectare. Isn't that small?'

'Yes, it is,' I agreed. 'Tell me about the people of Wayanad.'

'You see, before Wayanad fell into British hands, it was home to various tribal communities. In fact, there were hardly any non-tribal people here.'

'Which were the tribal communities that lived in Wayanad?'

'There were the Paniyas. In fact, a vast majority of the tribals in Kerala hail from the Paniya tribal sect. They are there in Wayanad in large numbers, and in the neighbouring parts of Kannur and Mallapuram districts.'

'Which are the other tribes in Wayanad?'

'The Adiyas, Kattunayakans, Kuruchiyans and Urali Kurumas.'

'Well, what happened after the British came to Wayanad?'

'Let me first tell you a story about how the British came in. A British engineer was the first one to come here with the help of a local tribal guide. He fell in love with what he saw here. Eager to take the credit for his magnificent discovery, he killed the tribal guide. But the soul of the tribal guide continued to haunt subsequent travellers to Wayanad. So a priest had to chain the turbulent spirit into a large ficus tree. This tree, bound by a prominent chain, exists even now. Let me tell you that the tree is a big tourist attraction here.'

'How interesting!'

'Well, after the British came, they opened up the plateau for cultivation of tea and other cash crops. They also built roads. So after the Second World War, there was an influx of settlers to Wayanad from all parts of Kerala and Karnataka. In the first year of their settlement, thousands of them died of malaria and attack by wild animals. Those who survived cleared the forest and transformed Wayanad to what it is now.'

We had reached Meppadi. First we went to see the AWS. It was very much like the one at Sulthan Bathery. We now went off to talk to Shri Raman Kutty.

Raman Kutty wore a white dhoti around the waist and a short-sleeved white shirt which, by the look of it, had seen better days. He had a big tilak on his head, made up of sandal paste. All in all, he looked like a man who had made his living with his hands.

'What do you do for a living?' I asked.

'I'm an agriculturist,' he said laconically. It was clear that he was a man of few words. I had to draw him out, I thought.

'How much land do you own?' I asked.

'About 5 hectares of land,' he said. 'But let me tell you that it is not very good land. Because it is on the other side of the mountain.'

'What are the crops that you grow?' I asked.

'I'll tell you. If the rain is good, I go for paddy. Otherwise, I grow ginger or some vegetables.'

'How do you get information about weather conditions?' Dr Manikiam asked.

'Like everyone here does; based on previous experience. I prepare my land very well before the monsoon rain comes. And start the operations when it rains. You see, we expect rain by the first week of June.'

'Do the rains come in the first week of June?' I asked.

'Sometimes. But there are years when the rain-gods are angry and the rains are delayed. Then I have a problem. We fully depend on rain. You see, I can't get water from the river which is 12 kilometres away.'

'Do you get weather information regularly?' Dr Manikiam asked.

'Well, I read the newspapers. I follow the weather forecasts given by the weather department. We also have a TV programme for the farmers. But do you know what my real problem is? I don't get the information I want.'

'What's the information you want?' Dr Manikiam asked.

'Well, I'd like to know when the first rain will come in June. After that, I want to know when we have the break conditions, when rain stops for even two weeks. Do you know why? Because I have to prepare for that to save my crops. The weather forecasts are too general and not useful for our village. So we have to consult some elders for advice.'

'Suppose we give you daily weather reports for your village such as temperature, rainfall, wind, etc.,' Dr Manikiam suggested, 'will it be useful?'

Raman Kutty nodded. 'Definitely. Such information will be useful for us to discuss and plan what to do. But we also want what'll happen during the next few days.'

'Yes,' I said, 'ISRO has already set up the Automatic Weather Station at Meppadi. That will look after your problem.'

'What'll this station do?' Raman Kutty asked.

It was Dr Manikiam who explained, 'It'll give you daily weather reports for Meppadi about temperature, rainfall and weather. It'll also give you weather forecasts for the next three to five days.'

'What kind of weather forecasts?' Raman Kutty asked.

'You see, forecast of rainfall, temperature and winds for three to five days,' Dr Manikiam explained.

'That'll be very useful,' he said. 'We'll be happy to use the machine to follow the weather. That way, I can avoid waste of fertilizers and pesticides if I can expect rains.'

'But let me tell you,' Dr Manikiam said with a note of caution in his voice, 'the weather forecast may not be fully correct presently. But it'll improve as we make more research studies.'

'That's all right,' Raman Kutty said with a laugh. 'Any information will be useful. Let me tell you that we also take such forecasts with a pinch of salt! But tell me something. Can you not tell us about large breaks in monsoon in advance? About the dry spells I mean?'

'We're going to do that,' Dr Manikiam said.

✳✳✳

Our next destination was Lakkidi.

'Who is the person we are going to meet in Lakkidi?' I asked.

'One Kochu Paniyan,' Dr Manikiam explained. 'He owns a coffee plantation. A small one, though.'

'Does he belong to the Paniya tribe you were talking about?'

Dr Manikiam looked thoughtful. 'Most likely,' he finally said. 'I haven't met him before. But his name would suggest that.'

'Tell me something about the Paniya tribe,' I suggested.

He immediately brightened up. It occurred to me that he might have missed his true vocation in life by becoming a weather scientist; he may have made a rather dedicated anthropologist.

'Paniyas were traditionally the bonded labourers,' he explained. 'They were sold along with the plantation by the landlords. They were also employed as professional coffee thieves by the higher castes.'

'Is that so?'

'Yes. Let me tell you a folk tale. It is about Pakkom Kota. That means the fort or prison of Pakkom. The story is about the escape of their ancestor from Pakkom Kota where they were treated as slaves. One day, they escaped from their feudal master.'

'How nice!' I said.

'They wandered into the forest, where they believed that their gods and goddesses would protect them. But the feudal master was able to send his god against their gods and they were threatened. The ancestors were thus caught again and enslaved. Even the existence of their gods was threatened by the feudal masters.'

'Do the Paniyas worship Hindu gods?'

'The Paniyas have only a crude idea of religion. Their major deity is called the Kali. They also worship the banyan tree. They don't cut such trees. If any Paniya does, he falls sick.'

We had reached Lakkidi. The mist rose as if to welcome us to this lovely place.

'This is a hill station,' Dr Manikiam announced. 'It has the highest rainfall in Kerala. You know something? There is mist formation in Lakkidi almost throughout the year.'

We went off to meet Kochu Paniyan. He wore a multi-coloured kaile mundu, wrapped around his waist and a loose shirt in very dark colours, the most prominent colour being red. He had a towel tied around his head like a headgear.

'What is your profession?' I asked.

'I am having a coffee plantation in Vythiri panchayat,' Kochu Paniyan said.

'What type of work do you have to do?' I asked.

'You see, we have to prepare *thadam*s in the plantation,' he explained.

'What are these thadams?' I asked.

'Thadams are the small ridges to save water.'

'Oh, I see,' I said. 'Anything else?'

'Yes, yes. We have to do many things. Like collecting the weeds. Spraying the pesticides.'

'How is the production in the plantation?' Dr Manikiam asked.

'Oh, it was very good earlier when good rains were there. Last two to three years, the rain is playing mischief. Now a new variety has come.'

'What is this new variety?' I asked.

'Vanilla. It brings lots and lots of money. But it needs very good rainfall and medium temperature. We are dependent on weather.'

'What type of weather information do you need?' Dr Manikiam asked.

'We need the latest information on daily rainfall. And also, the maximum and minimum temperatures. The production depends on this fully.'

'How do you get information on weather?' I asked. 'What are the sources, I mean?'

Kochu Paniyan shook his head. 'We are not getting much information. You see, we have to depend on the newspaper or TV to know when we can expect rain. But they aren't accurate.'

Dr Manikiam explained, 'You see, ISRO has set up this Automatic Weather Station at Noolpuzha which is very close to your place. It'll give you the daily data on weather. Will it be of any use to you?'

Kochu Paniyan nodded. 'Yes, yes. It will be of greatest use. We want rainfall and temperature data every day so that I can assess my production well in advance. Also, I'd like to know that there is less rain in Lakkidi or Meppadi.'

'Why Lakkidi or Meppadi?' I asked.

'They are the hilly regions here. Same as the place of my plantation in Vythiri. If I can get that information, I know my coffee will be in great demand.'

'You see,' Dr Manikiam explained, 'the AWS at Noolpuzha will give daily weather reports which will include the progress of monsoon. This will be a forecast for twenty-four hours. In addition, it will also give a weather forecast for the next three to five days which will include a forecast of rainfall, of temperature and winds. Will such information be useful to you?'

Kochu Paniyan laughed approvingly. 'If daily data is given, we know what to do,' he said. 'If the machine can tell when heavy rains are coming, we can prepare the thadams and protect our plants from flooding.'

※※※

It was time for us to leave. We were going to Kozhikode from Lakkidi. We crossed the Thamarassery Pass and the road descended down the Western Ghats through some mind-boggling bends and ridges.

We stopped for tea at a roadside stall. I remembered a question I had been meaning to ask Dr Manikiam. Lost in the seductive beauty of Wayanad, I had forgotten the question.

'I have a question to ask,' I told Dr Manikiam.

Dr Manikiam brightened. 'What is the question?' he asked with the enthusiasm of an eager student.

'You see, there have been such rapid developments in computer technology. And modelling nowadays is so sophisticated! Why is it that we still don't have reliable and accurate weather forecasts?'

Dr Manikiam said with an impish smile, 'In response, let me quote Patrick Young to you, "The trouble with weather forecasting is that it is right too often to be ignored and wrong too often to rely on it".'

It was funny, I thought, how weather has a way of inspiring witticisms.

13

IN THYAGARAJA'S LAND
The Village Resource Centre

Every year, in the month of December, poets and musicians from all over the country gather in large numbers at Tiruvaiyaru village in Tamil Nadu to sing in Thyagaraja's memory. Thyagaraja was a composer born in 1767 and is celebrated for the many Telugu songs he composed in praise of Lord Rama. His life and work are a source of great inspiration to young poets and musicians who believe they will be blessed with a melodious voice if they anoint the shrine with honey and sing Thyagaraja's songs at his memorial.

Tiruvaiyaru was the village where Thyagaraja spent the major part of his life and attained samadhi. Situated on the banks of the river Cauvery, Tiruvaiyaru is also known as the Panchananda Kshetra, was the abode of saints, poets and musicians, and of this place, Thyagaraja sang, 'the Panchananda Kshetra in the beautiful Chola country, nestling on the banks of the river Kaveri over which blows the gentle zephyr . . . where holy Brahmins chant the Vedas, a village to be coveted even by Lord Shiva.'

Thyagaraja was a composer with a difference. He extricated music from the tyrannical grip of words. In doing so, he composed for the common man. His compositions were such that they were easily accessible; almost anyone could understand his compositions and sing them. He was the composer with the common touch.

It is, perhaps, as a tribute to Thyagaraja, the common man's composer, that ISRO decided to locate its first ever Village Resource Centre in Tiruvaiyaru village. The Village Resource Centre is a facility established for the common villager so that he can have access to specific information on the natural resources of the village in local language in the local context for the overall development of the village. The Village Resource Centre provides multiple services to the villagers such as tele-agriculture services, telemedicine, tele-education, tele-health, tele-fisheries and other expert advice services for the benefit of the local people.

※※※

In Bangalore, I went across town to the Regional Remote Sensing Service Centre, a unit of ISRO, to talk to Shri Diwakar, the application scientist who did most of the work to establish the Village Resource Centre at Tiruvaiyaru. Diwakar is a lean, ascetic-looking man with a head so perfectly round that I sometimes get the impression that it was made in a toy factory. He is a man of relentless energy, and as an application scientist, his contribution to ISRO's mission of touching the common man's life with remote sensing applications has been immense.

We met in his rather small room, stuffed with official clutter. There were cupboards swollen with papers and files scattered on the floor. There were brightly coloured satellite imageries pinned to the walls that broke the tedium but made the room look rather surreal.

'Tell me about the Village Resource Centre,' I asked. 'What is it meant for?'

He explained, 'The idea is to carry the benefits of space-based systems to the ordinary villagers. What, in fact, drives the concept of the Village Resource Centre is that the people who live in the villages, often far-flung and in the interior, should have access to what ISRO has to offer.'

'Tell me, what does ISRO offer?'

'Several things; ISRO makes satellites and launches them. Two kinds of satellites—remote sensing and communication. These two kinds of satellites offer different kinds of benefits.'

'What kind of benefits?'

'You see, the communication technology is capable of providing benefits like telemedicine, tele-education, training facilities, developmental education and weather information. Remote sensing provides information on natural resources such as land, water, fisheries, etc. The idea is to create a single source in the village and through that, enable access for the villagers to the benefits that both remote sensing and communication technologies can provide.'

'So the Village Resource Centre is this single source at the village level?'

He nodded. 'Yes. We use the Village Resource Centre to deliver locale specific, useful information and services to the villages in both farm and non-farm sectors. In that sense, the Village Resource Centre is a unique programme.'

'Who runs the Village Resource Centre at the village level?' I asked. 'Is it ISRO?'

'What ISRO has done is to enlist the support of the NGOs. They look after the Village Resource Centre, run it and also mobilize the participation of the community in its activities.'

'Who is your partner NGO in the Village Resource Centre at Tiruvaiyaru?' I asked. 'And why did you decide to locate it at Tiruvaiyaru?'

'Well, our partner is the M.S. Swaminathan Research Foundation. The reason we decided to establish our first Village Resource Centre at Tiruvaiyaru was a very special one. You see, Tiruvaiyaru in Thanjavur district is located in a typical deltaic terrain with a challenge for groundwater recharge and optimal land use based on water availability.'

'What did you do to establish the Centre?'

'In the first instance, we had to set up both the communication and database infrastructure. The communication infrastructure consists of VSAT-based connectivity that supports audio, video and data connectivity. There is also the telemedicine facility with a digital ECG unit and related software, and a tele-education facility enabled through videoconferencing equipment.'

'What about the remote sensing part?'

'Yes. We had to establish that, too. We had to create a natural resource database including an action plan for optional land and water management with an easy-to-use software package. As a first step, we had to study the land use and land cover practices. For that, we had to procure and process multi-seasonal data from ISRO's remote sensing satellites.'

'What kind of processing did it involve?'

'You see, we had to geo-reference the satellite data to a common GIS base, and provide block and village boundaries, a cadastral database, infrastructure and settlements, drainage, an inventory of canals, river systems and surface water bodies, land use and land cover practices and soil maps, geomorphology, observation wells data and depth to the water table.'

'That's a long list,' I pointed out. 'But didn't you also have non-spatial elements in the database?'

'Yes. We had to obtain information in the form of socio-economic status of the village and link it to the village cadastral database. We also prepared a land and water resources action plan.'

'Did you have to carry out any fieldwork for the purpose?'

'Yes, of course. For preparing the natural resources database, we had to do extensive fieldwork. That was because we had to understand the problems that the people of the village faced, the limitations of the terrain, nature of the soil and availability of groundwater. We had to understand what the living conditions in the village were so that we could incorporate them into the database for working out suitable solutions for the benefit of the villagers.'

'Did you involve the villagers in this work?'

'Yes, we did, in a big way. We took up what is called the Participatory Rural Appraisal (PRA) exercise with the participation of the villagers. The villagers took part in the exercise and shared their socio-economic problems with us. The interaction with the individual farmers was captured in a database format for use under the information system.'

'What else?'

'We also took up an RRA exercise. That was to understand the overall socio-economic condition.'

'Tell me something,' I asked. 'Did you use the census data for all these activities?'

'Yes, we did. The census data, in fact, gave us an overall picture of the various socio-economic conditions and infrastructure at the village level. This helped us to concentrate on specific problems of a particular village.'

'Well, from what you say, the database appears to be very comprehensive,' I pointed out, 'but, tell me, is it useful to the villagers?'

'The answer is yes. Let me tell you that the database has proved to be of immense value to understand the modality of providing the right kind of services to the people in the villages.'

'In what way?'

'Let me give you an example. The socio-economic information in the form of the PRA and RRA has played a crucial role in understanding the living conditions and livelihood practices of the villagers. With that kind of background information, the Village Resource Centre coordinators at the village level are in a position to address all types of requirements of the villagers.'

'You said something about the easy-to-use software, didn't you?'

'Yes. We undertook an exercise on taking the GIS technology to the local people in the local language with local content and context.'

'That's very good,' I observed. 'But what exactly does it do?'

'This is an attempt to provide a customized solution. The idea is to incorporate all the information we talked about through a system that has the facility to query and display meaningful information for the benefit of the local people.'

'How is it user-friendly?'

'The package that we've provided is certainly user-friendly. We designed it in such a way that it is platform-independent. It can be installed on a simple desktop computer. And let me tell you that while designing and developing the system, we kept in mind the rural people and their requirements. The package provides a very simple mechanism to display and extract vital information on a wide variety of themes, both spatial and non-spatial.'

'Does the Village Resource Centre offer advisories to the villagers?'

'Oh, yes! That really is the thrust activity of the Village Resource Centre. The villagers use the centre to have interactive consultations, through videoconferencing, with the experts sitting at the M.S. Swaminathan Research Foundation at Chennai.'

'How many such interactions have taken place so far?'

Diwakar consulted some papers. 'About 1800 of them. Let me also give you the break-up: 22 per cent of these interactions have been by way of training, self-help groups account for 21 per cent of the interactions, agriculture advisories have been responsible for 15 per cent, and 9 per cent of the interactions have been for computer training.'

'What do these advisories consist of?'

'Let me cite the agriculture advisory as an example. We offer professional advisory on agriculture and cropping. This includes advice on crop, pest and disease management, seed and fertilizer availability, seed treatment, application of bio-fertilizers and bio-pesticides, and organic farming.'

'What about the marketability of the produce?' I asked. 'Have you factored that, too?'

'Yes, the Village Resource Centre provides advice on that. It includes information on price and marketability of the produce.'

'Anything on alternate cropping?'

'Yes, that's an important part of our advisory. The Village Resource Centre provides advice on *jathropa* cultivation, cultivation of medicinal plants, kitchen gardening and cultivation of newer varieties of banana.'

'You said something about training,' I pointed out. 'What kind of training does the centre provide on agriculture?'

'It provides training on several aspects. For example, on cultivation practices for *jathropa*, sweet sorghum, sugar beet, pulses, cultivation of oilseeds like gingelly, groundnut and sunflower, diseases of coconut, weed control, regulated market, farm pond construction, and harvesting rainwater.'

'That's very impressive,' I said. 'What about training on management of water resources?'

'The centre provides training on that, too. On aspects such as exploration of water through open and bore wells, location of farm ponds and irrigation tanks.'

'Well, the advisory on agriculture seems to be very comprehensive,' I commented. 'Does the centre provide advisories on other subjects as well?'

He nodded. 'Yes, yes. On livestock management, for example. The subjects that the centre covers are livestock varieties and rearing methods, marketing, diseases and their control such as deworming and vaccination. The centre also offers health advisories.'

'What does the health advisory offer advice on?'

'Generally on health awareness. The centre covers subjects like typhoid, hepatitis-B vaccination, and general health awareness on water-borne and viral diseases.'

'You mentioned something about computer training, didn't you?'

'Yes. That's an important advisory for the Village Resource Centre. It covers basics of computers and utilities for schoolchildren, and accounting software for the self-help groups. The centre also conducts training on income-generating activities and skill development.'

'That should be useful for the villagers,' I said. 'What are these activities?'

'For example, rice and banana produce, kitchen gardening and household poultry. Let me tell you that the Village Resource Centre gives an interesting advisory on farm finance and micro-credit.'

'What about telemedicine facilities?' I asked. 'Does the centre provide that?'

'Yes. The centre provides telemedicine connectivity. There is a telemedicine facility with a digital ECG unit and related software. The centre is connected to two super speciality hospitals—Sri Ramachandra Medical College and Research Institute, Chennai, and Aravind Eye Hospital, Madurai.'

'What about tele-education?'

'That's enabled through videoconferencing equipment at the Village Resource Centre. With the help of the Azim Premji Foundation fifty CDs have already been developed and these are being used. The centre is also linked with the ECO-Clubs in schools.'

※※※

Diwakar and I travelled to Tiruvaiyaru to meet the users of the Village Resource Centre and find out how they have benefited from it. The first person we met was Shri Ramalingam of Villiyanallor village which is about 7 kilometre from the centre.

'Are you an agriculturist?' I asked him.

He nodded. 'Yes. I have about 4.5 acres of land. I grow sugar cane and paddy.'

'But these crops need a lot of water, don't they?' I asked. 'Do you have access to that kind of water?'

He frowned. 'That's the problem. Of late, we have water problems in the Cauvery delta. That's why our usual cultivation practices have gone for a toss.'

'What are your usual cultivation practices?' I asked.

'Farmers here grow paddy crop twice a year. That's called kuruvai. In addition, we take a summer crop like gingelly or soyabean.'

'So what's the problem with that?' Diwakar asked.

He shook his head sorrowfully. 'No, that is not possible any more. So I'm looking out for meaningful cropping alternatives. I asked several persons. That's how I came to hear about ISRO'S Village Resource Centre at Tiruvaiyaru.'

'What did you do at the centre?' I asked.

'I started participating in all the meetings and training programmes conducted by the centre through video-conferencing.'

'Was that useful?' I asked.

'Let me tell you what was useful. I found the training provided by one Dr Muralidharan of the Tamil Nadu Agricultural University to be particularly useful. The training talked about cropping alternatives. Things like gingelly cultivation.'

'What about the database of the centre?' I asked. 'Did you try that for a recommendation on the cropping alternative?'

'Yes. The computer at the centre asked for the survey number of my land. I gave that. Do you know something? There was the picture of my land on the screen. And all the details of my land there on the screen in Tamil which I could read quite easily. Details like the kind of soil and water conditions. So I asked what should be the recommendation for a suitable crop for my land, given the shortage of water.'

'What was the recommended crop?' Diwakar asked.

'Groundnut. Then I was asked whether I would like to know what seed I should put. I said yes. The computer told me that it should be VRI-2. I noted that down. The computer also told me whom I should contact for the seeds. I was really impressed.'

'What about other things like fertilizers?' Diwakar asked.

'The computer gave very detailed recommendations about the fertilizer I should use and what should be the dosage.'

'Did you finally go for groundnut?' Diwakar asked.

'Yes, I did. But let me tell you that I kept coming back to the centre. You see, there's this package that the centre recommended. I can't get the name of that package. It's such a tongue-twister!'

'ICRISAT package,' Diwakar suggested.

Ramalingam nodded his head vigorously. 'Yes. That's the package. Do you know something? I followed the recommendations of the package given by the centre down to the last word.'

'Did you get a good crop?' I asked.

'Initially there was a problem. The pests attacked my groundnut crop. I came running to Tiruvaiyaru, to the centre. I was told what it was. Once again, I can't get the name.'

'It would have been Stemrot,' Diwakar suggested.

'Yes. That's what it was. The centre told me what to do. And I did accordingly. But let me tell you that after I did all that they told me at the centre, the groundnut crop has come up very well. I'm hoping to make enough money on this crop to compensate for all that I lost on the kuruvai.'

'Looks like you've benefited from the Village Resource Centre,' I said. 'Tell me something. Have you told this to other farmers you know?'

Ramalingam nodded. 'Yes. Do you know what they say about me in Villiyanallor? That, I have a big mouth. That I can't stop talking. I've told everyone how I've benefited from the Village Resource Centre.'

※※※

We met Shrimati Subha Gophi, a woman farmer from Tiruvaiyaru village, in the Village Resource Centre itself. I started talking to her.

'Do you come here often?' I asked.

She nodded. 'Yes, I'm a regular user of the Village Resource Centre.'

'What kind of advice are you looking for from the centre?' I asked.

'You see, I have about 2.5 acres of land. I now grow paddy and pulses. But they don't fetch me enough money. So what I'm looking for is agricultural technology for my farming.'

'Did the centre help you in finding that?' Diwakar asked.

'I've participated in several meetings and training programmes conducted here in the centre. I've learnt a lot. And combined with my experience of paddy cultivation over so many years, I have become an expert now.'

'That's very commendable,' I said.

'Isn't it?' she said smugly. 'Now, I am an adviser in the centre. The other day I talked to the farmers of Dindigul about paddy cultivation.'

I raised an eyebrow. 'How did you manage to do that?' I asked.

'It was in one of those interactive sessions. Through the videoconferencing here in the centre.'

'But what about your own problem?' I asked. 'Were you able to learn about the agricultural technology for farming in your own land?'

'Yes. I participated in the training programme that the centre conducted on gingelly cultivation. The trainer was Dr Muralidharan of Tamil Nadu Agricultural University. He has trained many farmers of this area. He is an expert on gingelly cultivation.'

'What was the training programme about?' I asked.

'We discussed some very important issues about gingelly cultivation. Such as weed control, pod borer attack, water management and micro-nutrient application.'

'Was that useful?' Diwakar asked.

'Yes, very useful. We were also given training on recommended dosages, marketing, hybrid seed availability, hybrid varieties, cultivation aspects of gingelly, and mechanism for testing the quality of gingelly.'

'That's very comprehensive, isn't it?' I suggested.

'Yes, it was comprehensive. But let me tell you what the best thing about the training programme was. We could ask as many questions as we wanted. And we could clear our doubts. Because, you see, it was an interactive training programme. I could clear all my doubts. And I learnt such a lot!'

'That's nice,' I observed.

'Another good thing happened at the training programme. The experienced gingelly cultivators who were there in the programme suggested that such training programmes should be conducted in the first week of March.'

'Why the first week of March?' I asked. 'What's so special about it?'

'It was because the training given should be timely, based on the climatic aspects,' she explained.

'That's true,' I said.

'Another interesting thing happened during the training programme. We had a meeting with traders in gingelly oil. That, I can tell you, was very useful from the marketing point of view.'

'So what did you finally decide?'

'Something useful. After participating in that training programme conducted by the Village Resource Centre, I decided that gingelly was the crop for me. I learnt all the agronomical practices of gingelly cultivation during that training programme.'

'Did you use that knowledge in your cultivation of gingelly?' Diwakar asked.

'Yes, I did,' she said. 'I took a summer crop of gingelly with very good results.'

'What were the results like?' I asked.

'Well, the yield I got was 750 kilos in a 2.5 acre plot. I'm very pleased with the yield.'

'Did you make a lot of money?' Diwakar asked.

She beamed. 'Yes, I did,' she said. 'And all that because of the Village Resource Centre.'

※※※

In the Village Resource Centre, we also met Mrs Kalaivani from Kallar Pasupathikovil village. She is from a traditional agricultural family and has studied up to SSLC. She is an enterprising woman farmer, who, along with some other women farmers, started a micro-enterprise to produce bio-central agents like Pseudomonas and Trychodama Vridhi on a small scale as an opportunity for self-employment. I started talking to Mrs Kalaivani.

'Let me congratulate you,' I said. 'I am told you've started a very successful micro-enterprise.'

'Thank you,' she said. 'Do you know what happened? I did that along with some of my other women farmer friends. Our micro-enterprise is on a very small scale. But it has provided opportunity for self-employment.'

'That's very creditable,' I commented. 'Let me ask you a question. How do you manage to market what you make?'

'That's where the Village Resource Centre has been extremely useful. After we make the bio-chemical agents, I bring them and keep them here in the Village Resource Centre.'

'Does that help?' I asked.

She nodded. 'It helps in input trading for the self-help groups. The Village Resource Centre also helps me in marketing my products for local buyers and traders.'

'Has the centre helped you in any other way?' Diwakar asked.

'Yes, it has. I am a regular visitor to the Information Centre here at the centre. There, I undergo training on the Accounting Systems for Self-help Groups.'

'What is this training?' I asked. 'Is this given by the centre?'

'They have developed this accounting software for the self-help groups. This is to assist the self-help groups working in the villages to keep their accounts properly and that too, on a regular basis.'

'But how does this training help the self-help groups of the villages?' Diwakar asked.

'It helps,' she said. 'Because, these self-help groups working at the village level take loans and financial assistance from the banks and the District Rural Development Agency of the Tamil Nadu government. So they have to submit monthly reports to the banks that have lent money to these groups. And also, to the District Rural Development Agency that has given them financial assistance. It is in the preparation and submission of these monthly reports that the training given by the centre helps.'

'Let me ask you a question,' I said. 'Has this training on the Accounting System helped you personally?'

She nodded. 'Yes, it has helped me greatly. I had problems maintaining accounts before. But after I went through this training, my accounts are in perfect order.'

'Are you now in a position to send monthly reports regularly and on time?' Diwakar asked.

'Yes,' she said. 'I now send my monthly reports to the bank and the District Rural Development Agency absolutely on time.'

'So it looks like the Village Resource Centre has been useful,' I said. 'Do you think the setting up of the centre was a good idea?'

She nodded. 'Yes,' she said. 'A very good idea. It has helped people in this area a great deal.'

※※※

We also met Arun Mozhi in the Village Resource Centre, Tiruvaiyaru.

'Do you come here very often?' I asked him.

He shook his head. 'No, this is my first visit to the Village Resource Centre. But I am a regular user of the services provided by the centre.'

I was puzzled. 'How is that possible?' I asked.

He laughed. 'I contact the Village Resource Centre regularly over the telephone. I ask questions on the agronomical practices to be followed under different conditions.'

'Do you find the services provided by the Village Resource Centre to be useful?' I asked.

'Yes, of course,' he said. 'Very useful. Let me give you an

example. I had this poor tillage in my paddy fields. I was getting desperate. I talked to many farmers about it. One of them advised me to get in touch with the Village Resource Centre and seek advice.'

'Did you do that?' I asked.

He nodded. 'Yes,' he said, 'I did. But over the telephone. I must say that the fellows manning the centre are very prompt and friendly. I requested the centre to advise me on how to improve the tillage in my paddy fields.'

'What was the advice from the centre?' I asked.

'I was asked to go in for urea and zinc sulphate,' he explained. 'The advice I got from the centre was very sound. It has given me very good results.'

'I'm happy for that,' I commented. 'Is there any other way in which the Village Resource Centre has proved useful to you?'

'Yes,' he said. 'I got a lot of help from the centre in respect of preparation of *panchakavya*.'

'What was the help?' I asked.

'Panchakavya is a growth regulator. It was prepared by me in my farm itself. It's used as a spray in my paddy fields. This has brought me good yield of paddy. And the grains are very good without any black spot.'

'But tell me, how does the centre come into this?' I asked.

'I was coming to that,' he said. 'I passed on this information to the centre. The centre, in turn, conveyed their recommendation to other farmers who regularly use the services offered by the centre. So what has happened now is that the growth regulator is widely used in our area, and with excellent results.'

'It's good to know that,' I said.

'A similar thing happened with vanilla cultivation,' he explained. 'There were many of my farmer friends who wanted to take up vanilla cultivation. They asked me for information. But I didn't have any, not being a vanilla cultivator myself. I told them to go to the Village Resource Centre, or like I do, contact the centre over the telephone.'

'Did they do that?' I asked.

'Yes, they did,' he said. 'And they were very happy with the kind of advice the centre gave them. Let me tell you about the kind of advice they got from the centre: specific advice on the market rate and crop suitability. And this was even before they decided to take up vanilla cultivation. The information given by the centre helped them to take a sound decision.'

'Let me ask you something,' I said. 'Do you think the Village Resource Centre has been useful to the farming community in Tiruvaiyaru and neighbouring villages?'

He nodded. 'Certainly yes. The Village Resource Centre has been very useful.'

※※※

We met Shri Suryanarayana in Varagur village. He is a farmer who owns about 1.5 acres of land.

'Tell me,' I asked. 'Do you use the services of the Village Resource Centre at Tiruvaiyaru?'

'Yes,' he said. 'I had a specific problem. Of course, many farmers in this area have a similar problem. I have this small plot of land and I used to grow paddy in kuruvai regularly.'

'And now, there is no release of water from Mettur,' I suggested.

'Oh, you already know that, don't you?' he said. 'Yes, there is no release of water. So what do I do? I need to grow an alternative crop which will consume less amount of water.'

'Did you approach the centre with your problem?' I asked.

He nodded. 'Yes, I did,' he said. 'You see, the centre has this unit called the Decision Support Centre. It's based on a particular kind of software.'

'The GeoVIS?' Diwakar suggested.

'Yes' he said, nodding his head. 'That's what it is. I posed my problem to the centre. And the advice given was that I should go in for the cumbu crop.'

'Did you take the advice given?' I asked.

'Yes, I did,' he said. 'But honestly, I didn't know what are the steps I should take because I was not familiar with the agricultural practices of the cumbu crop.'

'So what did you do?' Diwakar asked.

'Well, I queried the centre. They were very cooperative. They gave me all the technical guidance for the cultivation of the cumbu crop.'

'How was your cumbu crop?' I asked. 'Did you get good returns from it?'

He beamed. 'The crop was excellent. To tell you the truth, I made quite a bit of money.'

※※※

It was in Thirupalanam village that we met Revathi and Santhalakshmi. I was told that they were regular users of the services of the Village Resource Centre at Tiruvaiyaru.

'I believe,' I asked them, 'that you regularly use the services of the Village Resource Centre. Is that true?'

Both their heads nodded in unison. It was Revathi, however, who said, 'Yes. We do use the services of the centre quite a lot.'

'For what purpose?' I asked.

Revathi explained, 'We're engaged in training local volunteers, students and schoolchildren. So we approached the centre to give them computer training.'

'Did the centre do that?' I asked.

'Yes,' Santhalakshmi explained. 'A computer training programme was arranged for about fifty-two students. But that was not all! These students, in turn, were asked to train volunteers in their respective villages.'

'That's a great idea,' Diwakar commented. 'But tell us, did that really happen? Did these students who were trained in the centre go back and train people in their villages?'

'Yes, yes,' Santhalakshmi said. 'The students did train people in their villages. Something more important happened in the process.'

'What was the important thing?' I asked.

Santhalakshmi said, 'The computer training they got helped the villagers to access vital information of use from the local portal, Valani.'

'Valani?' Diwakar asked. 'What's that?'

It was Revathi who explained. 'It's an information system in Tamil developed by the Swaminathan Foundation for the benefit of the common people.'

※※※

We also met Shri Mohan in Thirupalanam village. He, we were told, was a regular user of the services offered by the Village Resource Centre; in fact, he has worked with the centre on behalf of the villagers of Thirupalanam to provide necessary information to them on the various training programmes conducted by the Village Resource Centre.

'You are a regular user of the Village Resource Centre, aren't you?' I asked.

'Yes,' he said. 'I use the services provided by the centre rather frequently, but that's not all! I also convey the advice offered by the centre to my fellow villagers.'

'Let me ask you something,' I said. 'Do you really think the services provided by the centre are useful to the villagers?'

He nodded. 'Yes. I'll give you an example. You see, in November 2004, there was an unusually heavy downpour in the Cauvery delta.'

I intervened to say, 'Wait, wait. Wasn't the Cauvery delta having a continuous dry spell for years? At least, that's what I remember from the newspapers.'

'That's true,' he said. 'There was a dry spell for three years continuously. That's why this sudden, heavy downpour in November 2004 caused such terrible havoc.'

'Why, what did it do?' Diwakar asked.

Mohan explained, 'It damaged the standing crop in the delta. Especially in the Tiruvaiyaru block. Do you know something? The Village Resource Centre stepped in to give advice to the farmers in this area.'

I intervened to point out, 'Wait a minute, would you? Was the centre established by that time?'

He nodded. 'Yes. Don't you remember? The prime minister inaugurated the Village Resource Centre in October 2004. So let us say the centre had just about come into existence.'

'Yes,' I said, 'there was a mix-up in my mind about the dates. Yes, what you say is right.'

He explained, 'The centre issued an advisory about what the farmers in the area should do. For good measure, the advisory was issued through the press and the All India Radio, Trichy.'

'What was the advisory like?' I asked.

'Well, it was in several parts. The first part was about the young samba crop affected by the rains. In that part of the advisory, the affected farmers were advised to apply 22 kilos of urea and 17 kilos of potash as top dress. That is, only after draining the excess water from the fields.'

'That's sound advice,' Diwakar suggested.

'Isn't it? That's what we thought. Do you know what the other things the advisory suggested were? It said that, wherever zinc deficiency was noted, the farmers should spray 1 kilo of zinc sulphate mixed with 2 kilos of urea in 200 litres of water.'

'But there is an easier alternative,' Diwakar said. 'Did the advisory suggest that?'

'If you mean the use of batteries,' Mohan said, 'yes, the advisory did suggest that. The advisory said that used batteries could be ground and sprinkled in the paddy fields or kept at the entrance of the irrigation channels.'

'Yes,' Diwakar said, 'that's what I had in mind. But let me tell you something. You seem to have a wonderful memory for numbers, don't you? You seem to remember all the numbers in the dosage prescribed in the advisory.'

He laughed. 'Well, it has got nothing to do with my memory, which is not as good as you make it out to be. You see, I had to note all the numbers of the dosage correctly because I had to use them in my own fields. You forget that

250 Touching Lives

I was also an affected farmer! And my added burden was that I had to advise other farmers too.'

'Was there anything else in the advisory?' I asked.

'Yes,' he said. 'You see, due to the heavy rains, the crop had already received atmospheric nitrogen. In all such places, use of urea could be avoided to reduce pest incidence.'

'But what about growth regulators?' Diwakar asked. 'The crop needed that, didn't it?'

'Yes. I was coming to that. You see, the advisory issued by the centre advised the farmers to spray panchakavya or amritakaraisal prepared out of the resources available with the farmers holding it.'

'That was sound advice,' Diwakar pointed out.

'I also think so,' Mohan agreed. 'The farmers were also advised to spray Pseudomonas at 400 grams per acre. This was to avoid incidence of diseases. The advisory recommended that the farmers shouldn't take up fresh planting of thaladi and late samba once again.'

'Why was that?' Diwakar asked.

'This was for a particular reason,' Mohan explained. 'You see, there was no guarantee of continued supply of water till the crop matured.'

'Yes,' I said. 'That is sound advice, considering that there is always that problem in the deltaic region of Tamil Nadu.'

'That was the reason for that particular recommendation,' Mohan explained. 'But the advisory of the centre gave them an alternative, too. It said the farmers could, instead, go in for groundnut, soyabean, pulses and gingelly. That, of course, depended on the soil conditions.'

'Tell me something,' I asked. 'Was this advice followed?'

Mohan nodded. 'Yes. It was followed in the new delta area.'

'Well, I find that the advisory issued by the Village Resource Centre was very comprehensive,' I said. 'But was it useful to the local farmers?'

'I can assure you that it was useful to all the farmers here in this area,' Mohan said. 'Let me tell you about my own case. I followed the advisory entirely. Without any exception. And it worked.'

'That's a nice thing to hear,' I said.

'Well, let me tell you something else,' Mohan said. 'I also passed on the information on the advisory to my fellow farmers in Thirupalanam village. They followed the advice scrupulously. And I can tell you that the results were positive for the entire village.'

'Thank you, Mohan,' Diwakar said effusively. 'We see that you have been an excellent emissary for the Village Resource Centre.'

He beamed us a radiant smile. 'Well, that's because the centre has been so useful to us. Useful for ordinary farmers like us.'

As we took leave of Mohan and proceeded to Tiruvaiyaru, I could see in my mind's eye Thyagaraja nodding in approval in his samadhi. He would have liked the idea of ISRO's Village Resource Centre. After all, the centre did touch the life of the common man. Didn't Thyagaraja do the same with his compositions?

INDEX

Abdul Nazir Sab State Institute of Rural Development, Mysore, 121
agricultural:
  advisory by Tiruvaiyaru Village Resource Centre, 236–37, 238–51;
  productivity improved, xiii; by watershed development, 130, 137–40, 153, 155–56;
  by weather forecast by Automatic Weather Station (AWS), 213, 217–19
Alakananda river, 165–66, 167–69, 170
Alirajpur, Jhabua, Madhya Pradesh, 131
Amrita Institute of Medical Sciences (AIMS), Kochi, 92–93, 94
aquifers, 128–29
Aravind Eye Hospital, Madurai, 94, 238
atmospheric processes, observation network, 214–15
Azim Premji Foundation, 238

Badi Dhami, Jhabua, Madhya Pradesh: watershed project, 152, 157
Badrinath, 170–71
Bakkhali, Sundarbans, 178–84
Bay of Bengal, 100, 103, 109
Bhils, 4–8, 11, 20, 131–32, 136–38, 142, 144, 154, 156;
  alcoholism, 7–8, 21–22; stopped, 151; criminals, 6; illiteracy, 17, 148–50; benefited from ISRO programme, 10–24;
  poverty, 4;
  superstitions, 7, 23;
  women, 18, 21
biodiversity, xiii
biogas plant, 143
biotic pressure, 175, 176
Bishop Richardson Hospital, Car Nicobar, 93
Boipariguda, Koraput, Orissa, 30–33

Index 253

Bori, Jhabua, Madhya Pradesh: watershed development, 129–43

canal system, 102
Canning, Sundarbans, 173–74, 180
Cauvery river, 231, 238, 248
Centre for Bio-Conservation and Development (CBCD), 152, 154, 156
Chamarajanagar, Karnataka: District Hospital, 95–96; educating children, 40*ff*; telemedicine connectivity, 95, 97–98
Chamoli, Madhya Pradesh: earthquake (1999) 169
check dam and reservoirs, 15, 129, 130, 131, 158
Chipko Movement, 167
Choti Malpur, Jhabua, Madhya Pradesh, 14–15
communication satellites, xii, 104, 233–34
Constitution of India, 73rd Amendment, 118
contour bunding, 153
coral reefs, 62
corruption, 1, 132, 134–36, 145–46, 158
cyclones, 103, 174

Dalei Ghai, 100–01, 112–14
Dashmantpur, Koraput, Orissa, 33–35
Data Relay Transponder (DRT), 216
deforestation, 102, 154, 166–69

Devi Deula, 110–12
Devi river, 100–1, 108–11, 114–16
Direct Reception System, 119
disaster management system, xii–xiii, 161–62
distance education programme in Chamarajanagar, Karnataka by ISRO, 41–59; in local vernacular, 46, 58
District Rural Development Agency, 243–44
drainage network, 163; surface drainage system, 191–93

Earth Observation System, 127
ecosystem, 103
educational awareness by ISRO programmes, 149–51
EDUSAT, 41–43, 49–51, 53, 59
embankments, 102–03, 104, 108, 183
Employment Assurance Scheme, 134
environmental awareness by ISRO programmes, 138, 144, 154–55, 179–88
environmental impact of development projects, xiii
exploitation of tribals, 34

fisheries:
advisory and information in local vernacular, 68, 70, 75; Catch per Unit Effort (CPUE), 76
fishing, biological activity, 66
flood(s):
forecasting, 103–04;

plain zoning, 103;
in hilly terrain, 165–66;
forest-based occupations, 36
Frazerganj, Sundarbans, 185–87

Ganga river, 161, 165, 172–76
GB Pant Hospital, Agartala, 82, 93
Geographical Information System (GIS), 234–35
geomorphology, 136, 234
geospatial framework, 104
Geo-Synchronous Satellite Launch Vehicle (GSLV), xii
GeoVIS, 246
grain bank, 133, 156–57
Gramkosh, 132–33, 140–42, 148, 155; *see also* self-help groups
groundwater maps, xiii
groundwater recharge, 128–29, 233–34

hazard zonation maps, 164–65, 169
health advisory by Tiruvaiyaru Village Resource Centre, 237
Hirakud dam/reservoir, 102, 107
Hooghly river, 173

ICRISAT, 240
Indian Meteorological Department (IMD), 215, 218–19
Indian National Centre for Ocean Information Services (INCOIS), 67–69, 73–75
Indira Awas Yojana, 134
Indira Gandhi Hospital, Kavaratti, 92

INSAT satellite, 119
INSAT 3-A, 216
Integrated Rural Development Programme (IRDP), 134
IRS-1A, xii

Jhabua, Madhya Pradesh;
Nahi panchayat, 134, 136, 140–42;
Moti panchayat, 134

Kalpana (satellite), 216
Kandhas, 30, 33, 35
Kedarnath, 166, 169
Kolyabada, Jhabua, Madhya Pradesh, 139, 140
Koraput, Orissa:
finding drinking water—an ISRO initiative, 25–39
Kotias, 33, 35
kuruvai, 238

Lakkidi, 219, 226–29
Lakshadweep:
fisheries, assistance by ISRO, 61–63, 76
Kavaratti, 60;
literacy, 65;
Minicoy, 64;
population, 64–65;
telemedicine connectivity, 92–93, 95
landslides, 160–62, 163–68;
causes, 162, 167–68;
induced floods, 165–66
literacy programme, Nellore district, 123

M.S. Swaminathan Research Foundation, Chennai, 233, 236, 248

Machhgaon, Orissa, 108–09
Machhlia village, Jhabua,
 Madhya Pradesh, 20–23
mahua liquor, 7–8
Makankavi Watershed
 Development, 143–52
Mandakini river, 166, 168
Mathalput, Koraput Orissa, 34–39
Meppadi, 219, 222;
 Automatic Weather Station
  (AWS), 224–26
Migration:
 in search of work, 31, 34;
 averted due to ISRO
  programmes, 140, 146,
  148, 150, 156
MYRADA, 138

Nanda Devi Biosphere Reserve, 170
Narayana Hrudayalaya,
 Bangalore, 81, 82, 95, 96–99
National Centre for Human
 Settlement & Environment
 (NCHSE), 143–45, 148
National Centre for Medium
 Range Weather Forecasting
 (NCMRWF), 217–19
National Drinking Water
 Technology Mission, 39
Neelkanth Peaks, 170–71
non-governmental organizations
 (NGOs), 120, 233;
 role in sodic land reclamation,
  196–97
Noolpuzha: Automatic Weather
 Station (AWS), 228–29

Ophthalmology, use of
 telemedicine connectivity, 93

Orissa:
 vulnerable to floods, 100–03;
 Remote Sensing and
  Application Centre, 26;
 Super Cyclone, (1999), 104–05, 112, 116
overgrazing, 168–69

Palya, Chamarajanagar,
 Karnataka: distance education
 programme by ISRO, 43–52
panchakavya, 245, 250
panchayats/panchayati raj
 system, 122;
 corruption, 132, 134–36,
  145–46, 158
Paniyas, 224, 227–29
Parjas, 30, 31, 33, 35
Participatory Rural Appraisal
 (PRA), 235
planning and policy formulation,
 xiii
Podu (shifting) cultivation,
 impact on mountain streams,
 27–28
Polar Satellite Launch Vehicle
 (PSLV), xii, 212
Potential Fishing Zone (PFZ),
 67–70, 73, 74–76
Prabha Eye Clinic, Bangalore, 94
Public Distribution Scheme
 (PDS), 124–25

Rabindranath Tagore
 International Institute of
 Cardiac Sciences (RTIICS),
 Kolkata, 80, 81–82, 84, 87, 88
rain water harvesting, 128–29

rainfall, responsible for landslides, 164, 167
Rajiv Gandhi National Drinking Water Mission, 39
Ralegaon Siddhi, 138–39
Regional Remote Sensing Centre, Bangalore, 232
Regional Remote Sensing Service Centre, Kharagpur, 173
relief and rehabilitation, 105, 162; *see also* disaster management system
remote sensing satellites, xii–xiii, 28–29, 104, 127, 129, 161–62, 190, 192–94, 233–34
reservation for women in panchayat bodies, 118–19
risk assessment framework, 162

Sanda village, Jhabua, Madhya Pradesh, 16–17
Saragur, Chamarajanagar, Karnataka: distance education programme by ISRO, 52–59
Satellite Instructional Television Experiment (SITE), 3
satellite-based development communication project, 4
training programmes, 4; training programme for teachers, 42, 50–51;
school education, 17–18
self-help groups of women:
in Alirajpur, 15, 17;
in Jhabua, Madhya Pradesh, 131, 133–34, 136, 141, 146–48, 150, 157;
in Gangetic plains, 197–98, 207–08;
in Tiruvaiyaru, 236–37, 243

Shankara Netralaya, Chennai, 94
Sitamma Footsteps, 122
slope morphology, role in landslide, 163
sodic soils, toxic to plants and injurious to human and animal health, 189
soil erosion, 102, 128, 130
soil stability information, 162
Space Application Centre (SAC), Ahmedabad, 218
Sri Ramachandra Medical College and Research Institute, Chennai, 93, 238
starvation deaths in Koraput, Orissa, 25
Sulthan Bathery, Automatic Weather Station (AWS), 219–22
Sundarbans:
agriculture, 177;
fisheries, 176, 177–78, 179–80;
livelihood, 177–78; tigers, 176–77
surface drainage system, *see* drainage network

Tada, Karnataka, 117–18
Tata Memorial Cancer Hospital, Mumbai, 94
technical interventions, 102
tele-education, xii, 233, 234; *see also* EDUSAT
telemedicine connectivity programme of ISRO, xii, 79–99, 233, 234, 238
thadams, 228, 229

Index

Thandla, Jhabua, Madhya Pradesh: Watershed Development Project, 152–59
tidal waves, 174–75, 183
Tiruvaiyaru, Tamil Nadu, 231; Village Resource Centre, 232–51;
Decision Support Centre, 246
Titly, 17
Training of Rural Youth for Self-employment (TRYSEM), 134
Tripura Sundari District Hospital, 77, 79–80, 96
tuna fishing, 63–64, 70–72

Udaipur, Tripura: telemedicine connectivity by ISRO, 77–92
Udaya TV, 48, 52, 56–57
Uttar Pradesh Sodic Land Reclamation Project (UPSLRP), 190–211

Valani (information system), 248
videoconferencing, 78, 81, 84, 119, 234, 236, 239
Village Resource Centre: Tiruvaiyaru, 232–51; Wayanad, 215, 218–19;

water/water resources management, 217, 219, 234, 237
watershed prioritization, 103
Wayanad Kerala: agricultural economy, 222–23; Automatic Weather Station (AWS), 216–20; people, 223–24
weather information, 212–29, 233
weathering, 163